Artificial Intelligence, Automation, and the Economy

Executive Office of the President

December 2016

Printed in the U.S.A.
ISBN-13: 978-1544643533

EXECUTIVE OFFICE OF THE PRESIDENT

WASHINGTON, D.C. 20502

December 20, 2016

Advances in Artificial Intelligence (AI) technology and related fields have opened up new markets and new opportunities for progress in critical areas such as health, education, energy, economic inclusion, social welfare, and the environment. In recent years, machines have surpassed humans in the performance of certain tasks related to intelligence, such as aspects of image recognition. Experts forecast that rapid progress in the field of specialized artificial intelligence will continue. Although it is unlikely that machines will exhibit broadly-applicable intelligence comparable to or exceeding that of humans in the next 20 years, it is to be expected that machines will continue to reach and exceed human performance on more and more tasks.

AI-driven automation will continue to create wealth and expand the American economy in the coming years, but, while many will benefit, that growth will not be costless and will be accompanied by changes in the skills that workers need to succeed in the economy, and structural changes in the economy. Aggressive policy action will be needed to help Americans who are disadvantaged by these changes and to ensure that the enormous benefits of AI and automation are developed by and available to all.

Following up on the Administration's previous report, *Preparing for the Future of Artificial Intelligence*, which was published in October 2016, this report further investigates the effects of AI-driven automation on the U.S. job market and economy, and outlines recommended policy responses.

This report was produced by a team from the Executive Office of the President including staff from the Council of Economic Advisers, Domestic Policy Council, National Economic Council, Office of Management and Budget, and Office of Science and Technology Policy. The analysis and recommendations included herein draw on insights learned over the course of the Future of AI Initiative, which was announced in May of 2016, and included Federal Government coordination efforts and cross-sector and public outreach on AI and related policy matters.

Beyond this report, more work remains, to further explore the policy implications of AI. Most notably, AI creates important opportunities in cyberdefense, and can improve systems to detect fraudulent transactions and messages.

Jason Furman
Chair, Council of Economic Advisers

John P. Holdren
Director, Office of Science and Technology Policy

Cecilia Muñoz
Director, Domestic Policy Council

Megan Smith
U.S. Chief Technology Officer

Jeffrey Zients
Director, National Economic Council

ARTIFICIAL INTELLIGENCE, AUTOMATION, AND THE ECONOMY

Contents

Executive Summary 1
    Economics of AI-Driven Automation 1
    Policy Responses 3
    Conclusion 4

Outreach and Development of this Report 5

Introduction 6

Economics of AI-Driven Automation 8
    AI and the Macroeconomy: Technology and Productivity Growth 8
    AI and the Labor Market: Diverse Potential Effects 10
    Historical Effects of Technical Change 11
    AI and the Labor Market: The Near Term 13
    What kind of jobs will AI create? 18
    Technology is Not Destiny—Institutions and Policies Are Critical 21

Policy Responses 26
    Strategy #1: Invest In and Develop AI for its Many Benefits 27
    Strategy #2: Educate and Train Americans for Jobs of the Future 30
    Strategy #3: Aid Workers in the Transition and Empower Workers to Ensure Broadly Shared Growth 34

Conclusion 43

References 44

ARTIFICIAL INTELLIGENCE, AUTOMATION, AND THE ECONOMY

Executive Summary

Accelerating artificial intelligence (AI) capabilities will enable automation of some tasks that have long required human labor.[1] These transformations will open up new opportunities for individuals, the economy, and society, but they have the potential to disrupt the current livelihoods of millions of Americans. Whether AI leads to unemployment and increases in inequality over the long-run depends not only on the technology itself but also on the institutions and policies that are in place. This report examines the expected impact of AI-driven automation on the economy, and describes broad strategies that could increase the benefits of AI and mitigate its costs.

Economics of AI-Driven Automation

Technological progress is the main driver of growth of GDP per capita, allowing output to increase faster than labor and capital. One of the main ways that technology increases productivity is by decreasing the number of labor hours needed to create a unit of output. Labor productivity increases generally translate into increases in average wages, giving workers the opportunity to cut back on work hours and to afford more goods and services. Living standards and leisure hours could both increase, although to the degree that inequality increases—as it has in recent decades—it offsets some of those gains.

AI should be welcomed for its potential economic benefits. Those economic benefits, however, will not necessarily be evenly distributed across society. For example, the 19th century was characterized by technological change that raised the productivity of lower-skilled workers relative to that of higher-skilled workers. Highly-skilled artisans who controlled and executed full production processes saw their livelihoods threatened by the rise of mass production technologies. Ultimately, many skilled crafts were replaced by the combination of machines and lower-skilled labor. Output per hour rose while inequality declined, driving up average living standards, but the labor of some high-skill workers was no longer as valuable in the market.

In contrast, technological change tended to work in a different direction throughout the late 20th century. The advent of computers and the Internet raised the relative productivity of higher-skilled workers. Routine-intensive occupations that focused on predictable, easily-programmable tasks—such as switchboard operators, filing clerks, travel agents, and assembly line workers—were particularly vulnerable to replacement by new technologies. Some occupations were virtually eliminated and demand for others reduced. Research suggests that technological innovation over this period increased the productivity of those engaged in abstract thinking, creative tasks, and problem-solving and was therefore at least partially responsible for the substantial growth in jobs employing such traits. Shifting demand towards more skilled labor raised the relative pay of this group, contributing to rising inequality. At the same time, a

---

[1] A more extensive introductory discussion of artificial intelligence, machine learning, and related policy topics can be found in the Administration's first report on this subject. *See* The White House, "Preparing for the Future of Artificial Intelligence," October 2016
(https://www.whitehouse.gov/sites/default/files/whitehouse_files/microsites/ostp/NSTC/preparing_for_the_future_of_ai.pdf).

slowdown in the rate of improvement in education, and institutional changes such as the reduction in unionization and decline in the minimum wage, also contributed to inequality—underscoring that technological changes do not uniquely determine outcomes.

Today, it may be challenging to predict exactly which jobs will be most immediately affected by AI-driven automation. Because AI is not a single technology, but rather a collection of technologies that are applied to specific tasks, the effects of AI will be felt unevenly through the economy. Some tasks will be more easily automated than others, and some jobs will be affected more than others—both negatively and positively. Some jobs may be automated away, while for others, AI-driven automation will make many workers more productive and increase demand for certain skills. Finally, new jobs are likely to be directly created in areas such as the development and supervision of AI as well as indirectly created in a range of areas throughout the economy as higher incomes lead to expanded demand.

Recent research suggests that the effects of AI on the labor market in the near term will continue the trend that computerization and communication innovations have driven in recent decades. Researchers' estimates on the scale of threatened jobs over the next decade or two range from 9 to 47 percent. For context, every 3 months about 6 percent of jobs in the economy are destroyed by shrinking or closing businesses, while a slightly larger percentage of jobs are added—resulting in rising employment and a roughly constant unemployment rate. The economy has repeatedly proven itself capable of handling this scale of change, although it would depend on how rapidly the changes happen and how concentrated the losses are in specific occupations that are hard to shift from.

Research consistently finds that the jobs that are threatened by automation are highly concentrated among lower-paid, lower-skilled, and less-educated workers. This means that automation will continue to put downward pressure on demand for this group, putting downward pressure on wages and upward pressure on inequality. In the longer-run, there may be different or larger effects. One possibility is superstar-biased technological change, where the benefits of technology accrue to an even smaller portion of society than just highly-skilled workers. The winner-take-most nature of information technology markets means that only a few may come to dominate markets. If labor productivity increases do not translate into wage increases, then the large economic gains brought about by AI could accrue to a select few. Instead of broadly shared prosperity for workers and consumers, this might push towards reduced competition and increased wealth inequality.

Historically and across countries, however, there has been a strong relationship between productivity and wages—and with more AI the most plausible outcome will be a combination of higher wages and more opportunities for leisure for a wide range of workers. But the degree that this materializes depends not just on the nature of technological change but importantly on the policy and institutional choices that are made about how to prepare workers for AI and to handle its impacts on the labor market.

Policy Responses

Technology is not destiny; economic incentives and public policy can play a significant role in shaping the direction and effects of technological change. Given appropriate attention and the right policy and institutional responses, advanced automation can be compatible with productivity, high levels of employment, and more broadly shared prosperity. In the past, the U.S. economy has adapted to new production patterns and maintained high levels of employment alongside rising productivity as more productive workers have had more incentive to work and more highly paid workers have spent more, supporting this work. But, some shocks have left a growing share of workers out of the labor force. This report advocates strategies to educate and prepare new workers to enter the workforce, cushion workers who lose jobs, keep them attached to the labor force, and combat inequality. Most of these strategies would be important regardless of AI-driven automation, but all take on even greater importance to the degree that AI is making major changes to the economy.

Strategy #1: Invest in and develop AI for its many benefits. If care is taken to responsibly maximize its development, AI will make important, positive contributions to aggregate productivity growth, and advances in AI technology hold incredible potential to help the United States stay on the cutting edge of innovation. Government has an important role to play in advancing the AI field by investing in research and development. Among the areas for advancement in AI are cyberdefense and the detection of fraudulent transactions and messages. In addition, the rapid growth of AI has also dramatically increased the need for people with relevant skills from all backgrounds to support and advance the field. Prioritizing diversity and inclusion in STEM fields and in the AI community specifically, in addition to other possible policy responses, is a key part in addressing potential barriers stemming from algorithmic bias. Competition from new and existing firms, and the development of sound pro-competition policies, will increasingly play an important role in the creation and adoption of new technologies and innovations related to AI.

Strategy #2: Educate and train Americans for jobs of the future. As AI changes the nature of work and the skills demanded by the labor market, American workers will need to be prepared with the education and training that can help them continue to succeed. Delivering this education and training will require significant investments. This starts with providing all children with access to high-quality early education so that all families can prepare their students for continued education, as well as investing in graduating all students from high school college- and career-ready, and ensuring that all Americans have access to affordable post-secondary education. Assisting U.S. workers in successfully navigating job transitions will also become increasingly important; this includes expanding the availability of job-driven training and opportunities for lifelong learning, as well as providing workers with improved guidance to navigate job transitions.

Strategy #3: Aid workers in the transition and empower workers to ensure broadly shared growth. Policymakers should ensure that workers and job seekers are both able to pursue the job opportunities for which they are best qualified and best positioned to ensure they receive an appropriate return for their work in the form of rising wages. This includes steps to modernize the social safety net, including exploring strengthening critical supports such as unemployment insurance, Medicaid, Supplemental Nutrition Assistance Program (SNAP), and Temporary Assistance for Needy Families (TANF), and putting in place new programs such as wage insurance and emergency aid for families in crisis. Worker empowerment also includes bolstering critical safeguards for workers and families in need, building a 21st century retirement system, and expanding healthcare access. Increasing wages, competition, and worker bargaining power, as well as modernizing tax policy and pursuing strategies to address differential geographic impact, will be important aspects of supporting workers and addressing concerns related to displacement amid shifts in the labor market.

Finally, if a significant proportion of Americans are affected in the short- and medium-term by AI-driven job displacements, policymakers will need to consider more robust interventions, such as further strengthening the unemployment insurance system and countervailing job creation strategies, to smooth the transition.

Conclusion

Responding to the economic effects of AI-driven automation will be a significant policy challenge for the next Administration and its successors. AI has already begun to transform the American workplace, change the types of jobs available, and reshape the skills that workers need in order to thrive. All Americans should have the opportunity to participate in addressing these challenges, whether as students, workers, managers, technical leaders, or simply as citizens with a voice in the policy debate.

AI raises many new policy questions, which should be continued topics for discussion and consideration by future Administrations, Congress, the private sector, academia, and the public. Continued engagement among government, industry, technical and policy experts, and the public should play an important role in moving the Nation toward policies that create broadly shared prosperity, unlock the creative potential of American companies and workers, and ensure America's continued leadership in the creation and use of AI.

Outreach and Development of this Report

This report was developed by a team in the Executive Office of the President including staff from the White House Council of Economic Advisers (CEA), Domestic Policy Council (DPC), National Economic Council (NEC), Office of Management and Budget (OMB), and Office of Science and Technology Policy (OSTP). This report follows a previous report published in October 2016 titled *Preparing for the Future of Artificial Intelligence* and the accompanying *National Artificial Intelligence Research and Development Strategic Plan*, developed by the National Science and Technology Council's (NSTC) Subcommittee on Machine Learning and Artificial Intelligence. This subcommittee was chartered in May 2016 by OSTP to foster interagency coordination and provide technical and policy advice on topics related to AI, and to monitor the development of AI technologies across industry, the research community, and the Federal Government. This report also follows a series of public-outreach activities as a part of the White House Future of Artificial Intelligence Initiative, designed to allow government officials to learn from experts and from the public, which included five co-hosted public workshops, and a public Request for Information (RFI).[2]

This report more deeply examines the impact of AI-driven automation on the economy and policy responses to it. It considers the economic evidence to better understand the lessons from past waves of automation, the impact already caused by the current wave of AI-driven automation and its prospects for the near future, and how AI-driven automation may affect workers in the future. The report also considers policy steps that are needed to address the economic dislocation caused by the arrival of these technologies and to prepare for longer-term trends in the economy caused by AI, automation, and other factors that are systemically disadvantaging certain workers. The report lays out three broad strategies for policymakers to consider.

---

[2] Ed Felten and Terah Lyons, "Public Input and Next Steps on the Future of Artificial Intelligence," *Medium*, September 6 2016 (https://medium.com/@USCTO/public-input-and-next-steps-on-the-future-of-artificialintelligence-458b82059fc3). Further details on the public workshops and the RFI can be found in the October 2016 report, *Preparing for the Future of Artificial Intelligence*.

Introduction

Recent progress in Artificial Intelligence (AI) has brought renewed attention to questions about automation driven by these advances and their impact on the economy. The current wave of progress and enthusiasm for AI began around 2010, driven by three mutually reinforcing factors: the availability of *big data* from sources including e-commerce, businesses, social media, science, and government;[3] which provided raw material for dramatically *improved machine learning approaches and algorithms;* which in turn relied on the capabilities of *more powerful computers.*[4] During this period, the pace of improvement surprised AI experts. For example, on a popular image recognition challenge that has a 5 percent human error rate according to one error measure,[5] the best AI result improved from a 26 percent error rate in 2011 to 3.5 percent in 2015. This progress may enable a range of workplace tasks that require image understanding to be automated, and will also enable new types of work and jobs. Progress on other AI challenges will drive similar economic changes.

Technical innovation has been expanding the American economy since the country's founding. American ingenuity has always been one of the Nation's greatest resources, a key driver of economic growth, and a source of strategic advantage for the United States. Remarkable homegrown innovations have improved quality of life, created jobs, broadened understanding of the world, and helped Americans approach their full potential. At the same time, they have forced Americans to adapt to changes in the workplace and the job market. These transformations have not always been comfortable, but in the long run—and supported by good public policy—they have provided great benefits.

The current wave of AI-driven automation may not be so different. For example, robots have made the economy more efficient. A 2015 study of robots in 17 countries found that they added an estimated 0.4 percentage point on average to those countries' annual GDP growth between 1993 and 2007, accounting for just over one-tenth of those countries' overall GDP growth during that time.[6] Some of that growth has been achieved by U.S. manufacturers adopting robots, allowing more goods to be produced while employing fewer workers at some facilities. AI in its many manifestations also holds promise to transform the basis of economic growth for countries across the world; a recent analysis of 12 developed economies (including the United States)

---

[3] "Big Data: Seizing Opportunities, Preserving Values," Executive Office of the President, May 2014, https://www.whitehouse.gov/sites/default/files/docs/big_data_privacy_report_may_1_2014.pdf.
[4] For more information about AI and its policy implications, see: The White House, "Preparing for the Future of Artificial Intelligence," October 2016.(https://www.whitehouse.gov/sites/default/files/whitehouse_files/microsites/ostp/NSTC/preparing_for_the_future_of_ai.pdf).
[5] The ImageNet Large Scale Visual Recognition Challenge provides a set of photographic images and asks for an accurate description of what is depicted in each image. Statistics in the text refer to the "classification error" metric in the "classification+localization with provided training data" task. *See* http://image-net.org/challenges/LSVRC/.
[6] Georg Graetz and Guy Michaels, "Robots at Work," *CEPR Discussion Paper No. DP10477,* March 2015 (http://papers.ssrn.com/sol3/papers.cfm?abstract_id=2575781).

found that AI has the potential to double annual economic growth rates in the countries analyzed by 2035.[7]

Some experts have characterized the rise of AI-driven automation as one of the most important economic and social developments in history. The World Economic Forum has characterized it as the lynchpin of a Fourth Industrial Revolution.[8] Furthermore, the economist Andrew McAfee wrote, "Digital technologies are doing for human brainpower what the steam engine and related technologies did for human muscle power during the Industrial Revolution. They're allowing us to overcome many limitations rapidly and to open up new frontiers with unprecedented speed. It's a very big deal. But how exactly it will play out is uncertain."[9]

At the same time, AI-driven automation has yet to have a quantitatively major impact on productivity growth. In fact, measured productivity growth over the last decade has slowed in almost every advanced economy. It is plausible, however, that the pace of measured productivity growth will pick up in the coming years. To the degree that AI-driven automation realizes its potential to drive tremendous positive advancement in diverse fields, it will make Americans better off on average. But, there is no guarantee that everyone will benefit. AI-driven changes in the job market in the United States will cause some workers to lose their jobs, even while creating new jobs elsewhere. The economic pain this causes will fall more heavily upon some than on others. Policymakers must consider what can be done to help those families and communities get back on their feet and assemble the tools they need to thrive in the transformed economy and share in its benefits.

---

[7] Paul Daugherty and Mark Purdy, "Why AI is the Future of Growth," 2016 (https://www.accenture.com/t20161031T154852__w__/us-en/_acnmedia/PDF-33/Accenture-Why-AI-is-the-Future-of-Growth.PDF#zoom=50).
[8] Klaus Schwab, "The Fourth Industrial Revolution: what it means, how to respond," World Economic Forum, January 2016 (https://www.weforum.org/agenda/2016/01/the-fourth-industrial-revolution-what-it-means-and-how-to-respond/). (The first three industrial revolutions are listed as those driven by steam power, electricity, and electronics.)
[9] Amy Bernstein and Anand Raman, "The Great Decoupling: An Interview with Erik Brynjolfsson and Andrew McAfee," *Harvard Business Review*, June 2015 (https://hbr.org/2015/06/the-great-decoupling).

Economics of AI-Driven Automation

Accelerating AI capabilities will enable automation of some tasks that have long required human labor. Rather than relying on closely-tailored rules explicitly crafted by programmers, modern AI programs can learn from patterns in whatever data they encounter and develop their own rules for how to interpret new information. This means that AI can solve problems and learn with very little human input. In addition, advances in robotics are expanding machines' abilities to interact with and shape the physical world. Combined, AI and robotics will give rise to smarter machines that can perform more sophisticated functions than ever before and erode some of the advantages that humans have exercised. This will permit automation of many tasks now performed by human workers and could change the shape of the labor market and human activity.

These transformations may open up new opportunities for individuals, the economy, and society, but they may also foreclose opportunities that are currently essential to the livelihoods of many Americans. This chapter explores the important role that AI-driven automation is likely to have in growing the economy and potential effects on labor markets and communities. It draws on economic theory and empirical studies of past technological transformations and applies these lessons to the current context. While there are many reasons to think that changes in the labor market prompted by AI-driven automation will be similar to what has been observed in the past, this chapter will also discuss arguments for how the current period could be different from previous technological revolutions.

Critically, technology alone will not determine the economic outcomes in terms of growth, inequality or employment. The advanced economies all have had access to similar levels of technology but have had very different outcomes along all of these dimensions because they have had different institutions and policies. But understanding the technological forces is critical to shaping the continued evolution of these policies.

AI and the Macroeconomy: Technology and Productivity Growth

To the extent that AI-driven automation resembles past forms of technological advancement, it will make important contributions to aggregate productivity growth.

For centuries, the American economy has adjusted to and evolved with technology. Many jobs that existed 150 years ago do not exist today, and jobs no one could have imagined then have taken their place. For example, in 1870, almost 50 percent of American employees worked in agriculture, supplying the Nation's food.[10] Today, thanks in large part to technological change, agriculture employs less than 2 percent of American workers and American food production exceeds domestic demand.[11] In this case, technological innovations, from McCormick harvesters

---

[10] Patricia A. Daly, "Agricultural Employment: Has the Decline Ended?" *Monthly Labor Review*, November 1981 (http://www.bls.gov/opub/mlr/1981/11/art2full.pdf).
[11] Bureau of Labor Statistics, "Employment Projections: Employment by major industry sector," December 2015 (http://www.bls.gov/emp/ep_table_201.htm).

to today's self-driving tractors, increased the productivity of the agricultural sector and contributed to increases in standard of living.[12]

One of the main ways that technology increases productivity is by decreasing the number of labor hours needed to create a unit of output. Labor productivity increases generally translate into increases in average wages, giving workers the opportunity to cut back on work hours and to afford more goods and services. Living standards and leisure hours could both increase, although to the degree that inequality increases—as it has in recent decades—it offsets some of those gains. The expectation that productivity increases would be accompanied by wage growth is what led John Maynard Keynes to predict in his 1930 essay on "Economic Possibilities for our Grandchildren" that, given the rates of technical progress, we might have achieved a 15-hour workweek by now.[13] While that prediction remains far off, over the last 65 years, most developed economies saw annual hours worked decline substantially (Figure 1). In the United States uniquely, however, this decline stopped in the late 1970s, and hours per worker has remained flat since then.

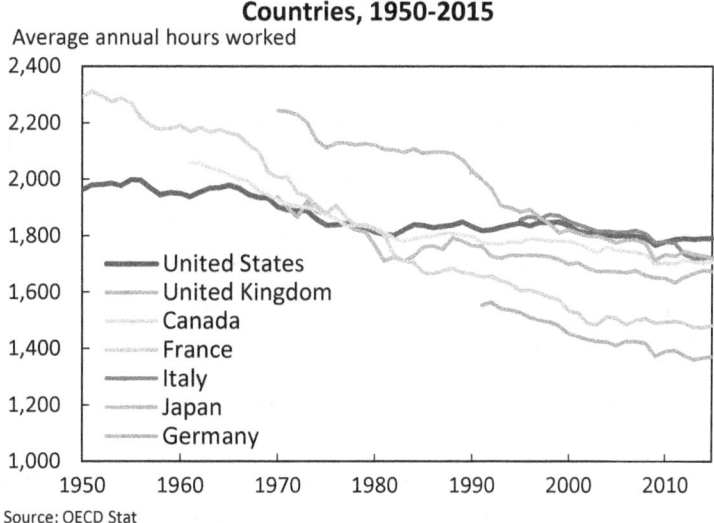

Figure 1: Average Annual Hours Worked per Worker, G-7 Countries, 1950-2015

Source: OECD Stat

Technology has been one of the main drivers of this productivity growth. Indeed, changes in technology help explain permanent productivity increases throughout the 1990s.[14] There is also

---

[12] USDA Economic Research Service, "Table 1. Indices of farm output, input, and total factor productivity for the United States, 1948-2013" (https://www.ers.usda.gov/data-products/agricultural-productivity-in-the-us/).
[13] John M. Keynes, "Economic Possibilities for our Grandchildren." In *Essays in Persuasion*, New York: W.W.Norton & Co., pp. 358-373, 1930. (http://www.econ.yale.edu/smith/econ116a/keynes1.pdf).
[14] Susanto Basu, John G. Fernald, and Matthew D. Shapiro, "Productivity growth in the 1990s: technology, utilization, or adjustment?" *Carnegie-Rochester Conference Series on Public Policy* 55(1): 117-65, 2001 (https://ideas.repec.org/a/eee/crcspp/v55y2001i1p117-165.html).

evidence that industrial robotic automation alone increased labor productivity growth by 0.36 percentage points across 17 countries between 1993 and 2007.[15]

The potential positive impact of AI-driven automation on productivity is particularly important given recent trends in productivity. In the last decade, despite technology's positive push, measured productivity growth has slowed in 30 of the 31 advanced economies, slowing in the United States from an average annual growth rate of 2.5 percent in the decade after 1995 to only 1.0 percent growth in the decade after 2005 (Figure 2). While a considerable amount of this slowdown in many countries—including the United States—is due to a slowdown in investment in capital stock, the slowdown in total factor productivity growth (the component influenced by technological change) has also been important.[16] This has contributed to slower growth in real wages and if continued will slow improvements in living standards.

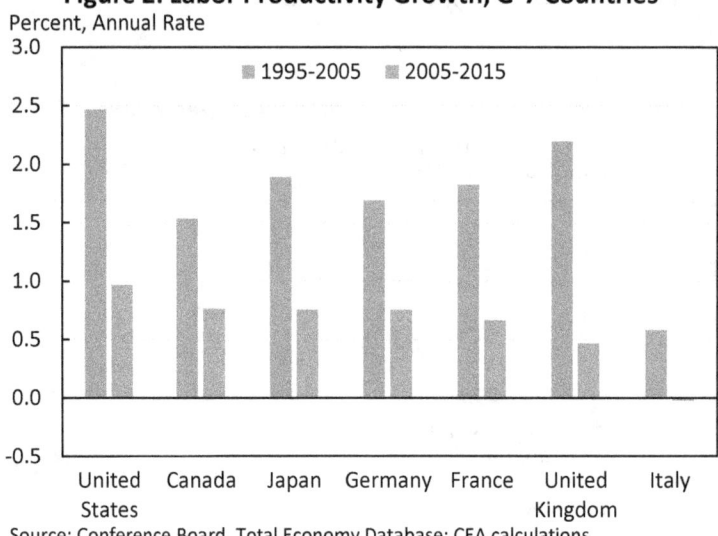

Figure 2: Labor Productivity Growth, G-7 Countries
Source: Conference Board, Total Economy Database; CEA calculations.

AI-driven automation could help boost total factor productivity growth and create new potential to improve the lives of Americans broadly. The benefits of technological change and economic growth, however, are not necessarily shared equally. This can depend on both the nature and speed of the technological change as well as the ability of workers to negotiate for the benefits of their increased productivity, as discussed below.

AI and the Labor Market: Diverse Potential Effects

Few would dispute that the industrial revolution largely made society better off, but the transition led to severe disruptions to the lives and communities of many agricultural workers, with

---

[15] Georg Graetz and Guy Michaels, "Robots at Work," *Centre for Economic Policy Research (CEPR) Discussion Paper No. DP10477*, 2015 (http://cep.lse.ac.uk/pubs/download/dp1335.pdf).
[16] Jason Furman, "Productivity Growth in the Advanced Economies: The Past, the Present, and Lessons for the Future," Speech at the Peterson Institute for International Economics, Washington, July 9 2015 (https://www.whitehouse.gov/sites/default/files/docs/20150709_productivity_advanced_economies_piie.pdf).

industrialization inducing many Americans to move to new communities where they could acquire new skills and put their time to new uses. Even during these periods of great technological change, America has maintained high employment. Over long periods, between 90 and 95 percent of the people in the United States who want a job at a given point in time can find one, and the unemployment rate is currently less than 5 percent.[17]

Historical Effects of Technical Change

Technological advances have historically had varied impacts on the labor market. New technologies may substitute for some skills while complementing others, and these trends change over time.[18] At times, new technologies have raised the productivity and increased employment opportunities for workers with little education, and other times for workers with more. To illustrate the diversity of potential impacts and provide a framework for understanding today, this section discusses historical examples of how innovations affected workers in different ways.

The 19th century was characterized by technological change that raised the productivity of lower-skilled workers and reduced the relative productivity of certain higher-skilled workers.[19] This kind of innovation has been called *unskill-biased technical change*. Highly-skilled artisans who controlled and executed full production processes saw their livelihoods threatened by the rise of mass production technologies that used assembly lines with interchangeable parts and lower-skilled workers. In reaction, some English textile weavers participated in the Luddite Riots of the early 1800s by destroying looms and machinery that threatened to undercut their highly-skilled, highly-paid jobs with lower-wage roles. Ultimately, the protesters' fears came true, and many skilled crafts were replaced by the combination of machines and lower-skill labor. There were also new opportunities for less-skilled workers and output per hour rose. As a result, average living standards could rise, but certain high-skill workers were no longer as valuable in the market.

Technological change tended to work in a different direction throughout the late 20th century. The advent of computers and the internet raised the relative productivity of higher-skilled workers, an example of *skill-biased technical change*. Routine-intensive occupations that focused on predictable, easily-programmable tasks—such as switchboard operators, filing clerks, travel agents, and assembly line workers—have been particularly vulnerable to replacement by new technologies.[20] Some entire occupations were virtually eliminated and demand for others reduced. Indeed, Nir Jaimovich and Henry Siu argue that the decline in manufacturing and other

---

[17] Bureau of Labor Statistics, Civilian Unemployment Rate, 1948-2016.
[18] Daron Acemoglu and David Autor, "Skills, tasks and technologies: Implications for employment and earnings," 2011, *Handbook of labor economics* 4(2011): 1043-171, (http://economics.mit.edu/files/5571).
[19] Daron Acemoglu, NBER reporter article, 2002; David Hounshell, *From the American system to mass production, 1800-1932: The development of manufacturing technology in the United States*, Baltimore: JHU Press, 1985; John A. James and Jonathan S. Skinner, "The Resolution of the Labor-Scarcity Paradox," *The Journal of Economic History*, 45(3): 513-40, 1985 (http://www.jstor.org/stable/2121750?seq=1#page_scan_tab_contents); Joel Mokyr, "Technological inertia in Economic History." *The Journal of Economic History*, 52(2): 325-38, 1992 (http://www.jstor.org/stable/2123111?seq=1#page_scan_tab_contents).
[20] David H. Autor, Frank Levy, and Richard J. Murnane, "The Skill Content of Recent Technological Change: An Empirical Exploration," *Quarterly Journal of Economics* 118(4): 1279-1333, 2003.

routine jobs is largely responsible for recent low labor demand for less educated workers.[21] Research suggests that technological innovation over this period increased the productivity of those engaged in abstract thinking, creative ability, and problem-solving skills and, therefore, is partially responsible for the substantial growth in jobs employing such traits.[22] Autor, Levy, and Murnane find that about 60 percent of the estimated relative demand shift favoring college-educated labor from 1970 to 1998 can be explained by the reduced labor input needed for routine manual tasks and the increased labor input for non-routine cognitive tasks, which tended to be more concentrated in higher-skilled occupations.[23] Given that college-educated labor was already more highly compensated, shifting demand towards college-educated labor and raising their relative pay contributed to rising income inequality.[24]

Like these past waves of technological advancements, AI-driven automation is setting off labor-market disruption and adjustment. Economic theory suggests that there must be gains from innovations, or they would not be adopted. Market forces alone, however, will not ensure that the financial benefits from innovations are broadly shared.

Market disruptions can be difficult to navigate for many. Recent empirical research highlights the costs of the adjustment process. In recent decades, U.S. workers who were displaced from their jobs—due to, for example, a plant closing or a company moving—experienced substantial earnings declines.[25] Autor, Dorn, and Hanson find that negative shocks to local economies can have substantial negative and long-lasting effects on unemployment, labor force participation, and wages.[26] Perhaps more significantly, over time, displaced workers' earnings recover only slowly and incompletely. Even ten or more years later, the earnings of these workers remain depressed by 10 percent or more relative to their previous wages.[27] These results suggest that for many displaced workers there appears to be a deterioration in their ability either to match their current skills to, or retrain for, new, in-demand jobs. AI-driven automation can act—and in some cases has already acted—as a shock to local labor markets that can initiate long-standing disruptions. Without some form of transfers and safety net, the economic benefits of AI-driven

---

[21] Nir Jaimovich and Henry E. Siu, "The Trend is the Cycle: Job Polarization and Jobless Recoveries." NBER Working Paper No. 18334, 2012 (http://www.nber.org/papers/w18334.pdf); Kerwin Kofi Charles, Erik Hurst, and Matthew J. Notowidigdo, "Housing Booms, Manufacturing Decline, and Labor Market Outcomes," Working Paper, 2016 (http://faculty.wcas.northwestern.edu/noto/research/CHN_manuf_decline_housing_booms_mar2016.pdf).

[22] Lawrence F. Katz and Kevin M. Murphy, "Changes in Relative Wages, 1963-1987: Supply and Demand Factors," *Quarterly Journal of Economics*, 107(Feb): 35-78, 1992; Daron Acemoglu, "Technical change, inequality, and the labor market," *Journal of economic literature* 40(1): 7-72, 2002; Autor, Levy, and Murnane 2003; Jaimovich and Siu 2012; David H. Autor and David Dorn, "The Growth of Low-Skill Service Jobs and the Polarization of the US Labor Market," *American Economic Review* 103(5): 1553-97, 2013 (http://www.ddorn.net/papers/Autor-Dorn-LowSkillServices-Polarization.pdf).

[23] Autor, Levy, and Murnane, 2003.

[24] Katz and Murphy, 1992.

[25] Steven J. Davis and Till Von Wachter, "Recessions and the Costs of Job Loss," Brookings Papers on Economic Activity, Economic Studies Program, The Brookings Institution, 43(2): 1-72, 2011 (http://www.econ.ucla.edu/workshops/papers/Monetary/Recessions%20and%20the%20Costs%20of%20Job%20Loss%20with%20Appendix.pdf).

[26] David H Autor, David Dorn, and Gordon H. Hanson, "The China Syndrome: Local Labor Market Effects of Import Competition in the United States," *American Economic Review* 103(6): 2121-68, 2013 (http://gps.ucsd.edu/_files/faculty/hanson/hanson_publication_it_china.pdf).

[27] Davis and Von Wachter, 2011.

automation may not be shared by all, and some workers, families, and communities may be persistently negatively affected.

AI and the Labor Market: The Near Term

Today, it may be challenging to predict exactly which jobs will be most immediately affected by AI-driven automation. Because AI is not a single technology, but rather a collection of technologies that are applied to specific tasks, the effects of AI will be felt unevenly through the economy. Some work tasks will be more easily automated than others, and some jobs will be affected more than others.

Some specific predictions are possible based on the current trajectory of AI technology. For example, driving jobs and housecleaning jobs are both jobs that require relatively less education to perform. Advancements in computer vision and related technologies have made the feasibility of fully automated vehicles (AVs), which do not require a human driver, appear more likely, potentially displacing some workers in driving-dominant professions. AVs rely upon, among other things, capabilities of navigating complex environments, analyzing dynamic surroundings, and optimization. Seemingly similar capabilities are required of a household-cleaning robot, for which the operational mandate is less specific (i.e. "clean the house," as opposed to the objective of navigating to a specific destination while following a set of given rules and preserving safety). And yet the technology that would enable a robot to navigate and clean a space as effectively as a human counterpart appears farther off. In the near to medium term, at least, drivers will probably be impacted more by automation than will housecleaners. The following sections lay out a framework for more general predictions of the effect of AI-driven automation on jobs.

Continued skill-biased technical change?

Recent research suggests that the effects of AI on the labor market in the decade ahead will continue the trend toward skill-biased change that computerization and communication innovations have driven in recent decades. Researchers differ on the possible magnitude of this effect. Carl Frey and Michael Osbourne asked a panel of experts on AI to classify occupations by how likely it is that foreseeable AI technologies could feasibly replace them over roughly the next decade or two.[28] Based on this assessment of the technical properties of AI, the relationship between those properties to existing occupations, and employment levels across occupations, they posit that 47 percent of U.S. jobs are at risk of being replaced by AI technologies and computerization in this period. Researchers at the Organisation for Economic Cooperation and Development (OECD), however, highlighted the point that automation targets tasks rather than occupations, which are themselves particular combinations of tasks.[29] Many occupations are likely to change as some of their associated tasks become automatable, so the OECD analysis

---

[28] Carl Frey and Michael Osborne, "The Future of Employment: How Susceptible are Jobs to Computerization," Oxford University, 2013
(http://www.oxfordmartin.ox.ac.uk/downloads/academic/The_Future_of_Employment.pdf).
[29] Melanie Arntz, Terry Gregory, and Ulrich Zierahn, "The Risk of Automation for Jobs in OECD Countries: A Comparative Analysis," OECD Social, Employment and Migration Working Papers No. 189, 2016
(http://www.oecd-ilibrary.org/docserver/download/5jlz9h56dvq7-en.pdf?expires=1480994298&id=id&accname=guest&checksum=6DC4B241A91EE860DC391585FF43C51C).

concludes that relatively few will be entirely automated away, estimating that only 9 percent of jobs are at risk of being completely displaced. If these estimates of threatened jobs translate into job displacement, millions of Americans will have their livelihoods significantly altered and potentially face considerable economic challenges in the short- and medium-term.

In addition to understanding the magnitude of the overall employment effects, it is also important to understand the distributional implications. CEA ranked occupations by wages and found that, according to the Frey and Osbourne analysis, 83 percent of jobs making less than $20 per hour would come under pressure from automation, as compared to 31 percent of jobs making between $20 and $40 per hour and 4 percent of jobs making above $40 per hour (Figure 3a). Furthermore, the OECD study estimates that less-educated workers are more likely to be replaced by automation than highly-educated ones (Figure 3b). Indeed, the OECD study's authors estimate that 44 percent of American workers with less than a high school degree hold jobs made up of highly-automatable tasks while 1 percent of people with a bachelor's degree or higher hold such a job. To the degree that education and wages are correlated with skills, this implies a large decline in demand for lower-skilled workers and little decline in demand for higher-skilled workers. These estimates suggest a continuation of skill-biased technical change in the near term.

Nevertheless, humans still maintain a comparative advantage over AI and robotics in many areas. While AI detects patterns and creates predictions, it still cannot replicate social or general intelligence, creativity, or human judgment. Of course, many of the occupations that use these types of skills are high-skilled occupations, and likely require higher levels of education. Further, given the current dexterity limits of the robotics that would be needed to implement mass AI-driven automation, occupations that require manual dexterity will also likely remain in demand in the near term.

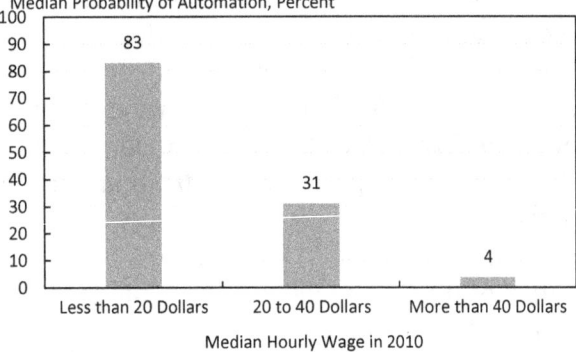

Figure 3a: Share of Jobs with High Probability of Automation, by Occupation's Median Hourly Wage

Source: Bureau of Labor Statistics; Frey and Osborne (2013); CEA calculations.

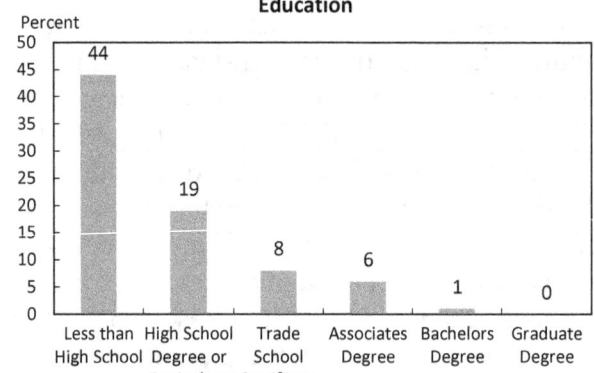

Figure 3b: Share of Jobs with Highly Automatable Skills, by Education

Source: Arntz, Gregory, and Zierahn (2016) calculations based on the PIAAC 2012.

Box 1: Automated Vehicle Case Study

A helpful case in understanding the types of effects AI may have on productivity and labor demand is the development of automated vehicles (AVs). Like other forms of technological disruption, AV technology will likely cause disruptions in the labor market as the economy adapts to new paradigms.

CEA estimates that 2.2 to 3.1 million existing part- and full-time U.S. jobs may be threatened or substantially altered by AV technology. Importantly, this is not a net calculation—it does not include the types of new jobs that may be developed—but rather a tally of currently held jobs that are likely to be affected by AI-enabled AV technology. A second caveat is that these changes may take years or decades to occur because there will be a further lag between technological possibility and widespread adoption.

This estimate of the number of current jobs likely displaced or substantially altered by AVs starts by identifying occupations that involve substantial driving and relatively few other responsibilities to lead and coordinate others, drawing occupation descriptions from the Bureau of Labor Statistics (BLS) and the Occupational Information Network (O*NET). For each of these occupations, an analysis of what non-driving tasks the occupation also requires yields an estimate of the share of jobs that will be displaced: driving jobs that also involve less-automatable tasks have a lower chance of disappearing. For example, the job of school-bus driver mixes both the tasks of driving and of attending to children. This job will not disappear, though it may evolve to focus heavily on the task of attending to children. As a result, AV technology may replace only a modest share of school-bus driver jobs, but child care workers will still be required. On the other hand, non-driving tasks are less important in inter-city bus driver jobs, and AV technology will likely replace a large share of these jobs.

Many jobs involve limited amounts of driving. These jobs are not included in the analysis below as individuals in these occupations would likely see a productivity boost, not a threat of displacement, as their time allocated to driving is freed up to focus on other critical tasks.

In addition to occupations identified from BLS and O*NET, CEA's estimate of jobs threatened or likely to be substantially altered by AVs also includes approximately 364,000 self-employed individuals driving either part- or full-time with ride-sharing services as of May 2015 that may find AV technology substituting for their services.

Table 1, below, presents the identified occupation categories.

Table 1

Occupation	BLS Code	# Total Jobs (BLS, May 2015)	Mean Hourly Wage	O*NET Occupation Description
Bus Drivers, Transit and Intercity	53-3021	168,620	$19.31	Drive bus or motor coach, including regular route operations, charters, and private carriage. May assist passengers with baggage. May collect fares or tickets.
Light Truck or Delivery Services Drivers	53-3033	826,510	$16.38	Drive a light vehicle, such as a truck or van, with a capacity of less than 26,000 pounds Gross Vehicle Weight (GVW), primarily to deliver or pick up merchandise or to deliver packages. May load and unload vehicle.
Heavy and Tractor-Trailer Truck Drivers	53-3032	1,678,280	$20.43	Drive a tractor-trailer combination or a truck with a capacity of at least 26,000 pounds Gross Vehicle Weight (GVW). May be required to unload truck. Requires commercial drivers' license.
Bus Drivers, School or Special Client	53-3022	505,560	$14.70	Transport students or special clients, such as the elderly or persons with disabilities. Ensure adherence to safety rules. May assist passengers in boarding or exiting.
Taxi Drivers and Chauffeurs	53-3041	180,960	$12.53	Drive automobiles, vans, or limousines to transport passengers. May occasionally carry cargo. Includes hearse drivers.
Self-employed Drivers	N/A	364,000[30]	$19.35[31]	N/A

As the above occupations involve other, critical non-driving tasks to varying degrees, Table 2, below, presents a weighted estimate of jobs threatened or substantially altered by AV technology in each occupation. For example, a weight of 0.60 indicates that approximately 60 percent of jobs in an occupation are threatened. These weights are generated by CEA using

---

[30] Calculations based on Hall and Krueger, 2016, (http://www.nber.org/papers/w22843.pdf), who note the number of active driver-partners for Uber more than doubled every six months in the period 2012-2014, and expect this growth rate to continue into 2015. CEA uses data from Harris and Krueger (2015) (http://www.hamiltonproject.org/assets/files/modernizing_labor_laws_for_twenty_first_century_work_krueger_harris.pdf) for only Uber and Lyft. Extrapolating to May 2015 with Hall and Krueger's method, CEA calculates approximately 324,000 individuals were active driver-partners with Uber in May 2015.
[31] Mean hourly earnings reported in Hall and Krueger (2016).

detailed job descriptions, case studies, and surveys of existing and planned technologies for each occupation.

Table 2

Occupation	# Total Jobs (BLS, May 2015)	Range of Replacement Weights	Range of # Jobs Threatened
Bus Drivers, Transit and Intercity	168,620	0.60 – 1.0	101,170 – 168,620
Light Truck or Delivery Services Drivers	826,510	0.20 – 0.60	165,300 – 495,910
Heavy and Tractor-Trailer Truck Drivers	1,678,280	0.80 – 1.0	1,342,620 – 1,678,280
Bus Drivers, School or Special Client	505,560	0.30 – 0.40	151,670 – 202,220
Taxi Drivers and Chauffeurs	180,960	0.60 – 1.0	108,580 – 180,960
Self-employed drivers	364,000	0.90 – 1.0	328,000 – 364,000
TOTAL JOBS	3,723,930		2,196,940 – 3,089,990

AV technology could enable some workers to focus time on other job responsibilities, boosting their productivity, and actually fostering wage growth among those still holding the reshaped jobs. For example, salespeople who currently spend a considerable amount of time driving could find themselves able to do other work while a car drives them from place to place, or inspectors and appraisers could fill out paperwork while their car drives itself. This should make these workers more productive, with AV technology serving as a complement, not a substitute. New jobs will also likely be created, both in existing occupations—cheaper transportation costs will lower prices and increase demand for goods and all the related occupations such as service and fulfillment—and in new occupations not currently foreseeable. Conversely, even occupations that have little connection to driving and divorced from the threat of automation could face pressure on wages due to an increased supply of similar, displaced workers.

In cases where downward pressure on wages or consolidation result in displacement of workers, private-sector solutions and public policy should aim to ensure smooth and quick transitions to new opportunities for these workers. CEA analysis finds that a share of workers in a few isolated occupations—truck drivers and delivery service drivers, in particular—currently enjoy a wage premium over others in the labor market with the same level of educational attainment. They may not be able to regain this wage premium if displaced

> without intervention to help them re-skill. Job search assistance, education, training and apprenticeships to build and certify new skills, and wage insurance could provide valuable support to them as they transition to finding new jobs.
>
> Policy responses to these challenges are discussed later in this report.

What kind of jobs will AI create?

Predicting future job growth is extremely difficult, as it depends on technologies that do not exist today and the multiple ways they may complement or substitute for existing human skills and jobs. To form an intelligent guess about what these jobs might be like, CEA has synthesized and extended existing research on jobs that would be directly created by AI. It is important to understand, however, that AI will also lead to substantial indirect job creation—to the degree it raises productivity and wages it may also lead to higher consumption that would support additional jobs across the economy in everything from high-end craft production to restaurants and retail.

CEA has identified four categories of jobs that might experience direct AI-driven growth in the future. Employment in areas where humans engage with existing AI technologies, develop new AI technologies, supervise AI technologies in practice, and facilitate societal shifts that accompany new AI technologies will likely grow. Current limits on manual dexterity of robots and constraints on generative intelligence and creativity of AI technologies likely mean that employment requiring manual dexterity, creativity, social interactions and intelligence, and general knowledge will thrive. Below are descriptions and potential examples of future employment for each category.

Engagement. Humans will likely be needed to actively engage with AI technologies throughout the process of completing a task. Many industry professionals refer to a large swath of AI technologies as "Augmented Intelligence," stressing the technology's role as assisting and expanding the productivity of individuals rather than replacing human work. Thus, based on the biased-technical change framework, demand for labor will likely increase the most in the areas where humans complement AI-automation technologies. For example, AI technology such as IBM's Watson may improve early detection of some cancers or other illnesses, but a human healthcare professional is needed to work with patients to understand and translate patients' symptoms, inform patients of treatment options, and guide patients through treatment plans. Shipping companies may also partner workers who pickup and deliver goods over the last 100 feet with AI-enabled autonomous vehicles that move workers efficiently from site to site. In such cases, AI augments what a human is able to do and allows individuals to either be more effective in their specialty task or to operate on a larger scale.

Development. In the initial stages of AI, development jobs are crucial and span multiple industries and skill levels. Most intuitively, there may be a great need for highly-skilled software developers and engineers to put these capacities into practical use in the world. To a certain extent, however, AI is only as good as the data behind it, so there will likely be increased demand for jobs in generating, collecting, and managing relevant data to feed into AI training processes. Applications of AI can range from high-skill tasks such as recognizing cancer in x-ray images to lower-skill tasks such as recognizing text in images. Finally, to an increasing degree, development may include those specializing in the liberal arts and social sciences, such as philosophers with frameworks for ethical evaluations and sociologists investigating the impact of technology on specific populations, who can give input as the new technologies grapple with more social complexities and moral dilemmas.

Supervision. This category encompasses all roles related to the monitoring, licensing, and repair of AI. For example, after the automated vehicle development phase, the need for human registration and testing of such technology to ensure safety and quality control on the roads will still likely exist. As a widespread new technology, AV will require regular repair and maintenance, which may expand mechanic and technician jobs in this space as well. Real-time supervision will also be required in exceptional, marginal, or high-stakes cases, especially those involving morality, ethics, and social intelligence that AI may lack. This might take the form of quality control of recommendations made by AI or online moderation when sensitive subjects are discussed. The capacity for AI-enabled machines to learn is one of the most exciting aspects of the technology, but it may also require supervision to ensure that AI does not diverge from originally intended uses. As machines get smarter and have improved ability to make practical predictions about the environment, the value of human judgement will increase because it will be the preferred way to resolve competing priorities.[32]

Response to Paradigm Shifts. The technological innovation surrounding AI will likely reshape features of built environment. In the case of AVs, dramatic shifts in the design of infrastructure and traffic laws—which are currently built with the safety and convenience of human drivers in mind—may be needed. The advent of self-driving cars may result in higher demand for urban planners and designers to create a new blueprint for the way the everyday travel landscape is built and used. Paradigm shifts in adjacent fields such as cybersecurity—demanding, for instance, new methods of detecting fraudulent transactions and messages—may also necessitate new occupations and more employment.

---

[32]Ajay Agrawal, Joshua Gans, and Avi Goldfarb, "The Simple Economics of Machine Intelligence," *Harvard Business Review,* November 17, 2016 (https://hbr.org/2016/11/the-simple-economics-of-machine-intelligence).

Box 2: The End of Work?

In addition to the arguments that AI and future technologies will broadly follow the same path as past technological revolutions, others make a more radical argument about the possible longer-run effects. They posit that AI could prove different from previous technological change because it has the ability to replicate something previously exclusive to humans: intelligence. There have long been fears that technology—the machines, the assembly lines, or the robots—would replace all human labor, but AI-driven automation has unique features that may allow it to replace substantial amounts of routine cognitive tasks in which humans previously maintained a stark comparative advantage. Initial waves of technology, such as the wheel and lever, allowed humans to do more by replacing or augmenting physical strength. Other processes allowed work to take place faster or more efficiently in a factory. Computers allowed calculations or pattern recognition to take place faster and augment humans' capacity to think or reason.

AI, though, may allow machines to operate without humans to such a degree that they fundamentally change the nature of production and work.[33] It may be that the question is no longer which segment of the population will technology complement, but whether the new technology will complement many humans at all, or if AI will substitute completely for much of human work. The skills in which humans have maintained a comparative advantage are likely to erode over time as AI and new technologies become more sophisticated. Some of this is evident today as AI becomes more capable at tasks such as language processing, translation, basic writing, or even music composition.

AI-driven technological change could lead to even larger disparities in income between capital owners and labor. For example, Brynjolfsson and McAfee argue that current trends in the labor market, such as declining wages in the face of rising productivity, are indicative of a more drastic change in the distribution of economic benefits to come. Rather than everyone receiving at least some of the benefit, the vast majority of that value will go to a very small portion of the population: "superstar-biased technological change." Superstar-biased technological change is somewhat similar to skill-biased technological change, but the benefits of technology accrue to an even smaller portion of society than just the highly-skilled workers. The winner-take-most and winner-take-all nature of the information technology market means that the fortunate few are likely to emerge as victors of the market. This would exacerbate the current trend in the rising fraction of total income going to the top 0.01 percent (Figure 4).

---

[33] Erik Brynjolfsson and Andrew McAfee, *The Second Machine Age: Work, Progress, and Prosperity in a Time of Brilliant Technologies,* WW Norton & Company, 2014.

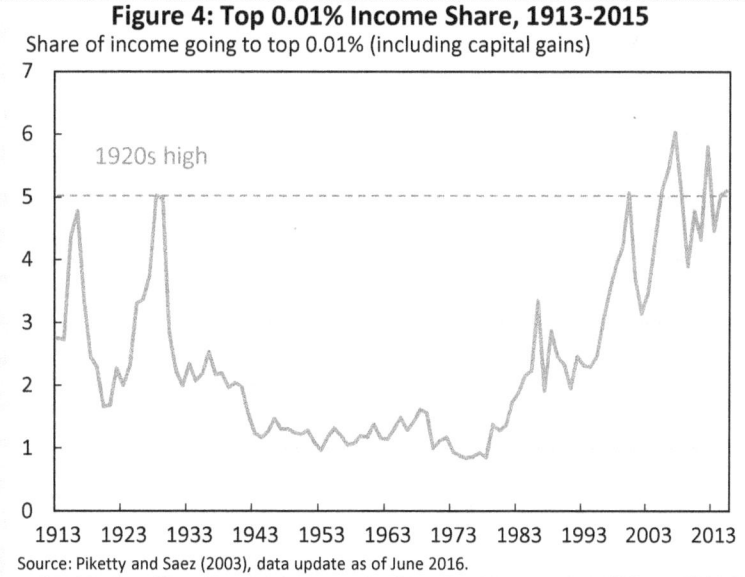

Figure 4: Top 0.01% Income Share, 1913-2015
Share of income going to top 0.01% (including capital gains)
Source: Piketty and Saez (2003), data update as of June 2016.

In theory, AI-driven automation might involve more than temporary disruptions in labor markets and drastically reduce the need for workers. If there is no need for extensive human labor in the production process, society as a whole may need to find an alternative approach to resource allocation other than compensation for labor, requiring a fundamental shift in the way economies are organized.

Although this scenario is speculative, it is included in this report to foster discussion and shed light on the role and value of work in the economy and society. Ultimately, AI may develop in the same way as the technologies before it, creating new products and new jobs such that the bulk of individuals will be employed as they are today.[34]

Technology is Not Destiny—Institutions and Policies Are Critical

A key determinant of how AI-induced technological change will affect people in the future is the ability of workers to extract the benefits of their increased productivity. For decades after World War II, the share of income going to the bottom 90 percent of workers was roughly unchanged. But since the late 1970s, the bottom 90 percent of households have seen their income fall from two-thirds of the total to about one half of the total share of U.S. income. For much of this period, moreover, productivity growth did not translate to higher real wages for low-income and even middle-income American workers.

This reduced share of income is partly the result of the fact that labor compensation is being increasingly unevenly distributed. But since 2000, it is also because the distribution of benefits going to capital and labor have also been diverging. Starting in about 2000, corporate profits as a share of GDP (a measure of the capital share of GDP) started to increase and labor share of GDP

---

[34] For more discussion, see John W. Budd, *The Thought of Work*, Cornell University Press, 2011.

began decreasing (Figure 6). The labor share of GDP reached a historical low, though it has trended up somewhat over the last 2 years.

Figure 5: Non-Farm Labor and Corporate Profits Share of GDP, 1950-2016

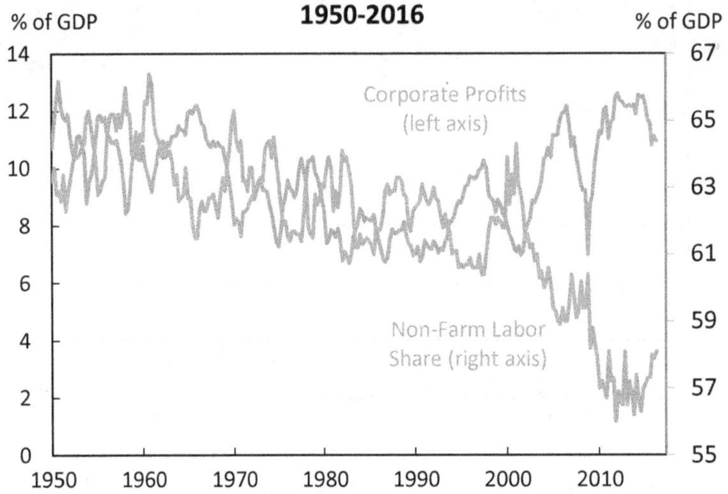

Source: Bureau of Economic Analysis, Bureau of Labor Statistics; CEA calculations

How AI and AI-driven automation will shape the distribution of gains in coming years depends on non-technical factors including aspects of both the broader economy and policy institutions. First, the direction of innovation is not a random shock to the economy but the product of decisions made by firms, governments, and individuals. Economic factors can drive the direction of technological change. Second, there is a role for policy to help amplify the best effects of automation and temper the worst.

Technological advancement is generated and adopted into the economy as the product of choices of entrepreneurs, workers, and firms looking to better serve a market or streamline a production process, in the context created by public investments in basic and applied research, infrastructure, and other public goods. In a process of directed technical change, incentives draw investment towards more potentially profitable innovations and so the types of technological change that are likely to occur, among those which are technologically feasible, are those which are most profitable.[35] Research examining firms' decisions to innovate argues that the tendency towards unskill-biased technical change in the 1800s came about because it was profitable to create technologies that replaced expensive and scarce resources (skilled artisans) with relatively cheap and abundant resources (machines and low-skilled workers).

In contrast, research suggests that skill-biased technical change of the 1990s was a function of increases in the supply of educated workers, which made innovations that raised their productivity more profitable because they could be used widely.[36] The Frey and Osborne study

---

[35] Daron Acemoglu, "Directed Technological Change, *Review of Economic Studies* 69(4): 781-809, 2003 (http://restud.oxfordjournals.org/content/69/4/781.short).
[36] Acemoglu, 2003.

and the OECD study suggest that this trend of skill-biased technical change may continue with AI, as the most automatable occupations tend to be low-wage and low-skill.

On the other hand, the wage premium for higher-skilled labor has increased over time (Figure 6), which creates a countervailing incentive to invest in innovations that might raise the productivity of lower-skill, lower-price workers. For example, automation technology that embodies expertise in medical imaging hardware and software allows middle-skill personnel to make medical diagnosis, reducing demand for more-expensive specialists. Similarly, the introduction of proprietary tax preparation software has allowed less-skilled tax preparers to replace certified accountants in some situations. In both cases, the demand for high-skill computer programmers also increases slightly but their work diffuses widely and scales cheaply. Thus technological change does not happen in a vacuum. The trajectory of AI may shift and change depending on non-technical, competitive incentives.

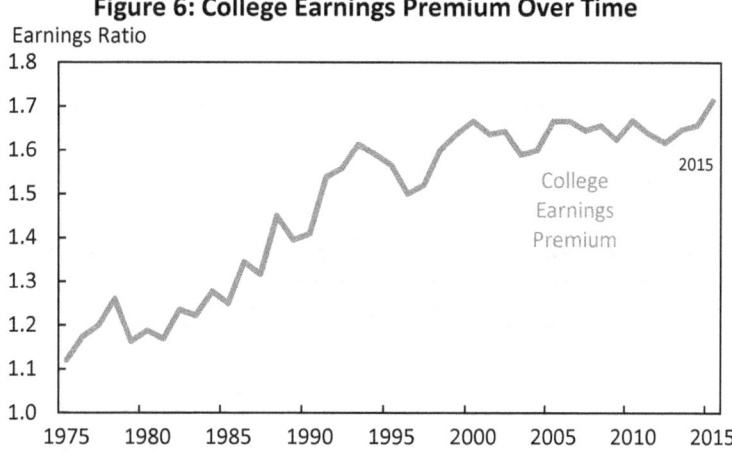

Policy plays a large role in shaping the effects of technological change. Therefore, even if Frey and Osborne's predictions that almost 50 percent of occupations are threatened by new automation technologies are accurate, the labor market impacts also depend on a country's institutions and policies. While relative wages depend on the demand for different levels of skill, which is partially a function of technology, wages also depend on the supply of labor at various skill levels, which is influenced by the distribution of educational opportunity and attainment.[37] Relative wages also depend on collective bargaining,[38] minimum wage laws, and other institutions and policies that affect wage setting.

Over the last 4 decades, other major advanced countries have experienced technological changes similar to the United States, yet the United States has seen both a greater increase in income inequality and higher overall levels of inequality, as shown in Figure 7. While most other

---

[37] Claudia Goldin and Lawrence F. Katz, *The Race between Education and Technology*, 2008.
[38] Bruce Western and Jake Rosenfeld, "Unions, norms, and the rise in US wage inequality," *American Sociological Review* 76(4): 513-37, 2011.

advanced economies have seen declines in prime-age male labor force participation, moreover, the decline in the United States has been steeper than in almost every other advanced economy, as shown in Figure 8.

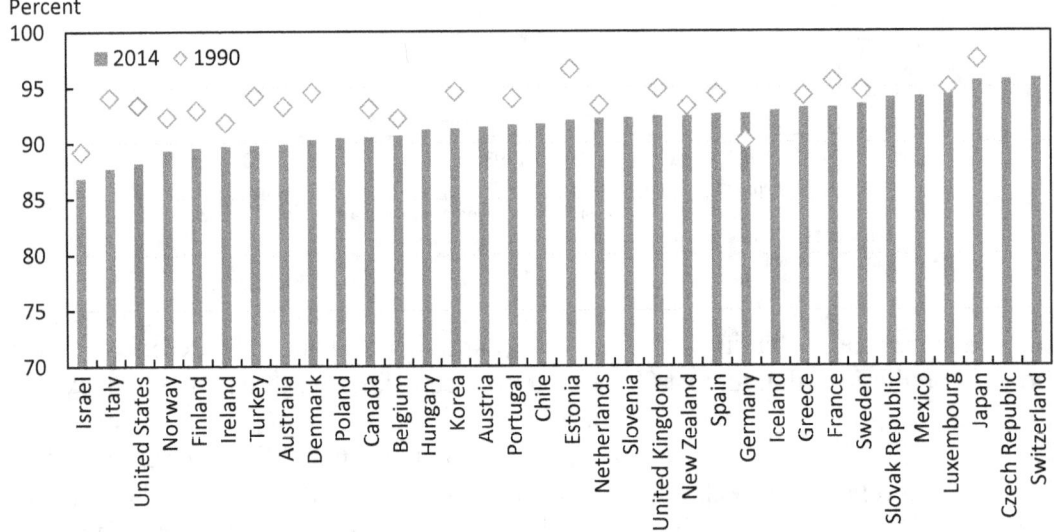

Figure 7: Share of Income Earned by Top 1 Percent, 1975-2015

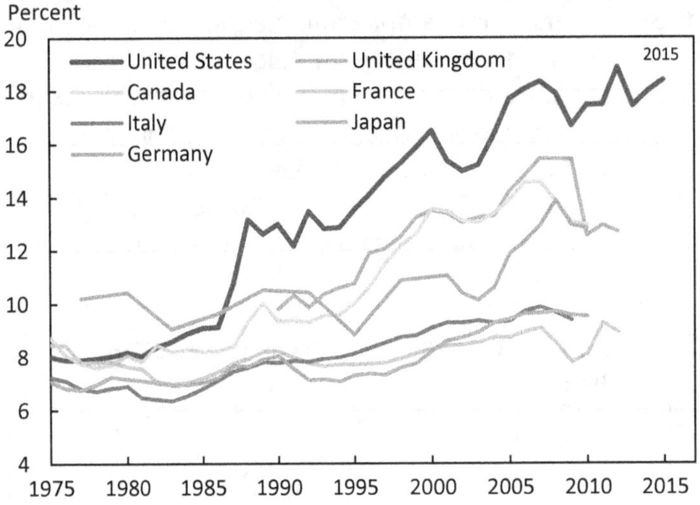

Source: World Wealth and Income Database.

Figure 8: Prime-Age Male Labor Force Participation Rates Across the OECD

If the difference between the United States and the other countries in inequality and labor force participation pictured above cannot be explained by the rate of technological change and which tasks are automatable, then it suggests that differences in a country's policies and institutions may mediate these changes. For example, other countries tend to invest far more resources on active labor market programs that help workers navigate job transitions, such as training programs and job-search assistance. While OECD member countries outside of the United States spent, on average, 0.6 percent of GDP on active labor market policies in 2014, spending by the

United States was just 0.1 percent of GDP (Figure 9). The United States, moreover, now spends less than half of what it did on such programs 30 years ago as a share of GDP (Figure 10).[39]

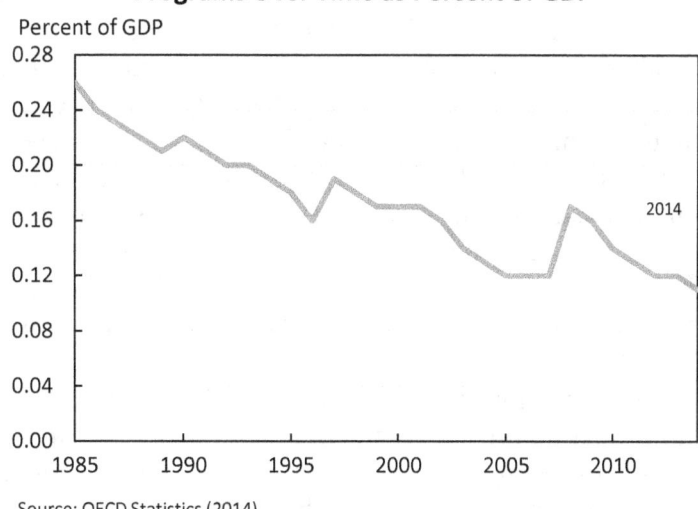

Figure 9: Public Expenditure on Active Labor Market Programs (% of GDP)

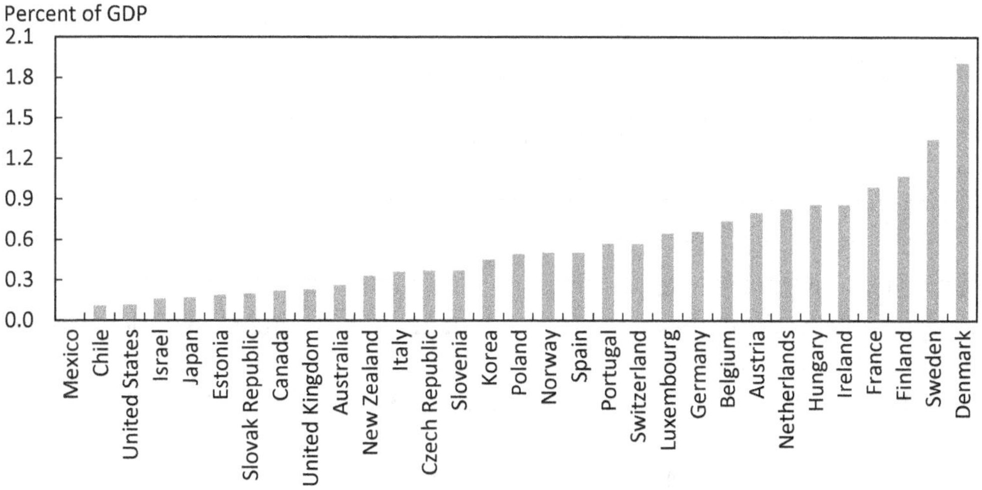

Note: Data for Ireland, Poland, and Spain from 2013; Data for UK for 2011.
Source: OECD Statistics (2014)

Figure 10: U.S. Public Expenditure on Active Labor Market Programs Over Time as Percent of GDP

Source: OECD Statistics (2014)

It is possible for the economy to generate high levels of employment with more advanced automation and higher levels of productivity than there are today. But Federal policies will need to play a role in helping Americans to navigate transitions in labor market demand caused by changes in technology over time.

---

[39] OECD, "Labour market programmes: expenditure and participants", *OECD Employment and Labour Market Statistics* (database), 2016 (http://stats.oecd.org/viewhtml.aspx?datasetcode=LMPEXP&lang=en#).

Policy Responses

AI-driven automation stands to transform the economy over the coming years and decades. The challenge for policymakers will be to update, strengthen, and adapt policies to respond to the economic effects of AI.

Although it is difficult to predict these economic effects precisely with a high degree of confidence, the economic analysis in the previous chapter suggests that policymakers should prepare for five primary economic effects:

- Positive contributions to aggregate productivity growth;
- Changes in the skills demanded by the job market, including greater demand for higher-level technical skills;
- Uneven distribution of impact, across sectors, wage levels, education levels, job types, and locations;
- Churning of the job market as some jobs disappear while others are created; and
- The loss of jobs for some workers in the short-run, and possibly longer depending on policy responses.

There is substantial uncertainty about how strongly these effects will be felt, and how rapidly they will arrive. It is possible that AI will not have large, new effects on the economy, such that the coming years are subject to the same basic workforce trends seen in recent decades—some which are positive, and others which are worrisome and may require policy changes. At the other end of the range of possibilities, the economy might potentially experience a larger shock, with accelerating changes in the job market, and significantly more workers in need of assistance and retraining as their skills are no longer valued in the job market. Given presently available evidence, it is not possible to make specific predictions, so policymakers must be prepared for a range of potential outcomes. At a minimum, some occupations such as drivers and cashiers are likely to face displacement from or restructuring of their current jobs, leading millions of Americans to experience economic hardship in the short-run absent new policies.

Because the effects of AI-driven automation will likely be felt across the whole economy, and the areas of greatest impact may be difficult to predict, policy responses must be targeted to the whole economy. In addition, the economic effects of AI-driven automation may be difficult to separate from those of other factors such as other technological changes, globalization, reduction in market competition and worker bargaining power, and the effects of past public policy choices. Even if it is not possible to determine how much of the current transformation of the economy is caused by each of these factors, the policy challenges raised by the disruptions remain, and require a broad policy response.

In the cases where it is possible to direct mitigations to particular affected places and sectors, those approaches should be pursued. But more generally, this report suggests and discusses below three broad strategies for addressing the impacts of AI-driven automation across the whole U.S. economy:

1. Invest in and develop AI for its many benefits;
2. Educate and train Americans for jobs of the future; and
3. Aid workers in the transition and empower workers to ensure broadly shared growth.

Strategy #1: Invest In and Develop AI for its Many Benefits

If care is taken to responsibly maximize its development, AI will make important, positive contributions to aggregate productivity growth, and advances in AI technology hold incredible potential to help America stay on the cutting edge of innovation. Indeed, CEA Chair Jason Furman has written that his biggest worry about AI is "that we do not have enough [of it]."[40] AI technology itself has opened up new markets and new opportunities for progress in critical areas such as health, education, energy, economic inclusion, social welfare, transportation, and the environment. Substantial innovation in AI, robotics, and related technology areas has taken place over the last decade, but the United States will need a much faster pace of innovation in these areas to significantly advance productivity growth going forward.[41] With the right investment in AI and policies to support a larger and more diverse AI workforce, the United States has the potential to accelerate productivity and maintain the strategic advantages that result from American leadership in AI.[42]

Invest in AI research and development

Government has an important role to play in advancing the AI field by investing in research and development. Throughout the public outreach on AI conducted by OSTP, government officials heard calls from business leaders, technologists, and economists for greater government investment in AI research and development. Leading researchers in AI were optimistic about sustaining the recent rapid progress in AI and its use in an ever wider range of applications. A strong case can be made in favor of increased Federal funding for research in AI.

The Administration published its *Artificial Intelligence Research and Development Strategic Plan* in October 2016, laying out a detailed strategy and roadmap for government-funded AI research and development.

Develop AI for cyberdefense and fraud detection

Currently, designing and operating secure systems requires a large investment of time and attention from experts. Automating this expert work, partially or entirely, may enable strong security across a much broader range of systems and applications at dramatically lower cost, and

---

[40] Jason Furman, "Is This Time Different? The Opportunities and Challenges of Artificial Intelligence." Remarks at AI Now: The Social and Economic Implications of Artificial Intelligence Technologies in the Near Term. New York University, New York, July 7, 2016.
(https://www.whitehouse.gov/sites/default/files/page/files/20160707_cea_ai_furman.pdf).
[41] *Ibid.*
[42] More detailed policy recommendations for investments in AI research and development, and development of the AI workforce, see the previous Administration report, *Preparing for the Future of Artificial Intelligence* (https://www.whitehouse.gov/sites/default/files/whitehouse_files/microsites/ostp/NSTC/preparing_for_the_future_of_ai.pdf), and the accompanying *Artificial Intelligence Research and Development Strategic Plan* (https://www.nitrd.gov/PUBS/national_ai_rd_strategic_plan.pdf).

may increase the agility of cyber defenses. Using AI may help maintain the rapid response required to detect and react to the landscape of ever evolving cyber threats. There are many opportunities for AI and specifically machine-learning systems to help cope with the sheer complexity of cyberspace and support effective human decision making in response to cyberattacks.

Future AI systems could perform predictive analytics to anticipate cyberattacks by generating dynamic threat models from available data sources that are voluminous, ever-changing, and often incomplete. These data include the topology and state of network nodes, links, equipment, architecture, protocols, and networks. AI may be the most effective approach to interpreting these data, proactively identifying vulnerabilities, and taking action to prevent or mitigate future attacks. Results to-date in DARPA's Cyber Grand Challenge competition demonstrate the potential of this approach.[43] The Cyber Grand Challenge was designed to accelerate the development of advanced, autonomous systems that can detect, evaluate, and patch software vulnerabilities before adversaries have a chance to exploit them. The Cyber Grand Challenge Final Event was held on August 4, 2016. To fuel follow-on research and parallel competition, all of the code produced by the automated systems during the Cyber Grand Challenge Final Event has been released as open source to allow others to reverse engineer it and learn from it.

AI also has important applications in detecting fraudulent transactions and messages. AI is widely used in the industry to detect fraudulent financial transactions and unauthorized attempts to log in to systems by impersonating a user. AI is used to filter email messages to flag spam, attempted cyberattacks, or otherwise unwanted messages. Search engines have worked for years to maintain the quality of search results by finding relevant features of documents and actions, and developing advanced algorithms to detect and demote content that appears to be unwanted or dangerous. In all of these areas, companies regularly update their methods to counter new tactics used by attackers and coordination among attackers.

Companies could develop AI-based methods to detect fraudulent transactions and messages in other settings, enabling their users to experience a higher-quality information environment. Further research is needed to understand the most effective means of doing this.

Develop a larger, more diverse AI workforce

The rapid growth of AI has dramatically increased the need for people with relevant skills to support and advance the field. The AI workforce includes AI *researchers* who drive fundamental advances in AI and related fields, a larger number of *specialists* who refine AI methods for specific applications, and a great number of *users* who operate those applications in specific settings. For researchers, AI training is inherently interdisciplinary, often requiring a strong background in computer science, statistics, mathematical logic, and information theory.[44] For

---

[43] Cyber Grand Challenge (https://www.cybergrandchallenge.com).
[44] There is also a need for the development of a strong research community in fields outside of technical disciplines related to AI, to examine the impacts and implications of AI on economics, social science, health, and other areas of research.

specialists, training typically requires a background in software engineering as well as in the application area. For users, familiarity with AI technologies is needed to apply them reliably.

All sectors face the challenge of how to diversify the AI workforce. The lack of gender and racial diversity in the AI-specific workforce mirrors the significant and problematic lack of diversity in the technology industry and the field of computer science more generally. Unlocking the full potential of the American people, especially in entrepreneurship, as well as science, technology, engineering, and mathematics (STEM) fields, has been a priority of the Administration. The importance of including individuals from diverse backgrounds, experiences, and identities, especially women and members of racial and ethnic groups traditionally underrepresented in STEM, is one of the most critical and high-priority challenges for computer science and AI.

Research has shown that diverse groups are more effective at problem solving than homogeneous groups, and policies that promote diversity and inclusion will enhance our ability to draw from the broadest possible pool of talent, solve our toughest challenges, maximize employee engagement and innovation, and lead by example by setting a high standard for providing access to opportunity to all segments of our society.[45]

Policymakers also will need to address new potential barriers stemming from any algorithmic bias. Firms are beginning to use consumer data, including data sets collected by "third-party" data services companies, to determine individuals' fitness for credit, insurance and even employment. Other firms are pioneering the use of games, simulations, and electronic tests to determine the "fit" of job applicants to a team. To do this, automated algorithms may be trained with data about successful team members in order to look for applicants that resemble them. One important benefit of these innovations is enabling companies to recruit and hire candidates based on demonstrated skills and abilities rather than pedigree, which will become even more critical as people gain skills on the job. But if such training sets are based on a less diverse current workforce, the biases of the existing group may be built into the resulting decisions and may unfairly exclude new potential talent. While employment laws and the Fair Credit Reporting Act currently impose certain restrictions on the use of credit history in making employment decisions, additional statutory or regulatory protections may be need to be explored in this space.

The previous report, *Preparing for the Future of Artificial Intelligence*, discusses workforce needs, including the strong case for increasing diversity, and lays out a detailed plan for AI workforce development.[46]

Support market competition

Competition from new and existing firms has always played an important role in the creation and adoption of new technologies and innovations, and this is no different in the case of AI. Startups

---

[45] Aparna Joshi and Hyntak Roh, "The Role of Context in Work Team Diversity Research: A Meta-Analytic Review," *Academy of Management Journal* 52: 599-627, 2009; Vivian Hunt, Dennis Layton, and Sara Prince, "Why Diversity Matters," McKinsey & Company, 2015 (http://www.mckinsey.com/business-functions/organization/our-insights/why-diversity-matters).

[46] The White House, *Preparing for the Future of Artificial Intelligence*, (https://www.whitehouse.gov/sites/default/files/whitehouse_files/microsites/ostp/NSTC/preparing_for_the_future_of_ai.pdf).

are a critical pathway for the commercialization of innovative new ideas and products. Startups, or the possibility of entry by a startup, also incentivize established firms to innovate and reduce costs. Competition pushes firms to invest in new technologies that help to lower costs, and also to invest in innovations that can lead to improvements in the quality of existing products.

The rapid evolution of technology can pose challenges for developing sound pro-competition policies, both in terms of defining the scope of the market as well as assessing the degree of contestability or the possibilities for disruption. For example, while it is probably too early to assess the role of AI in competition policy, one might imagine that when a large incumbent has access to most of the customer data in the market, it is able to use AI to refine its products better than any potential entrant could hope, and can thereby effectively foreclose entry.

Strategy #2: Educate and Train Americans for Jobs of the Future

As AI changes the nature of work and the skills demanded by the labor market, American workers will need to be prepared with the education and training that can help them continue to succeed. If the United States fails to improve at educating children and retraining adults with the skills needed in an increasingly AI-driven economy, the country risks leaving millions of Americans behind and losing its position as the global economic leader.

The United States led the world in economic gains from the industrial revolution in part due to major investments in its workforce. In the 20th century, America shifted from a mostly agrarian economy to an industrial economy. During this period, as today, the type of work and skills required to do work underwent a major transformation. To meet the needs of the new economy, the United States rapidly expanded access to education through high school, and by 1930 America was far ahead of European countries in terms of widely available, free, and publically-provided secondary education.[47] The average American born in 1951 had 6.2 more years of schooling than an American born in 1876. These increases in schooling led to tangible economic gains: economists estimate that educational attainment explains 14 percent of annual increases in labor productivity during that period.[48] Americans used education to adapt to new jobs which could not even have been imagined in prior decades. Today's AI-driven transformation will likely require similar realignments. College- and career-ready skills in math, reading, computer science, and critical thinking are likely to be among the factors in helping workers successfully navigate through unpredictable changes in the future labor market. Providing the opportunity to obtain those skills will be a critical component of preparing children for success in the future.[49]

Educate youth for success in the future job market

A key step towards preparing individuals for the economy of the future is providing quality education opportunities for all.

---

[47] Ajay Chaudry, "The Case for Early Education in the Emerging Economy," *Roosevelt Institute-The New American Economy's Learning* Series, August 2016, p. 3 (http://rooseveltinstitute.org/wp-content/uploads/2016/08/The-Case-for-Early-Education-in-the-Emerging-Economy.pdf).
[48] Claudia Dale Goldin and Lawrence Katz, *The Race between Education and Technology*, 2008.
[49] The White House, Economic Report of the President 2016, Chapter 4.

While, in the past, many jobs paying decent wages could be done with low levels of skill, continuing changes in technology, including AI, will make such jobs less common in the future. While it is unclear exactly how progress in AI and other technologies will affect the jobs of the future, policymakers must address the low levels of proficiency in basic math and reading for millions of Americans. Despite strong progress over the past 8 years, the United States is still falling behind rather than leading the world in key dimensions for successfully navigating this transition. Children from low-income families start kindergarten over 1 year behind peers in language skills.[50] Student performance in mathematics in China is on average 2 years above U.S. students.[51] American students from lower socioeconomic backgrounds score 15 percent lower on international assessments than higher income peers.[52] And year after year, a stubborn gap persists between how well white students are doing compared to their African American and Latino classmates.[53]

For all students, coursework in STEM, and specifically in areas such as computer science, will likely be especially relevant to work and citizenship in an increasingly AI-driven world. To respond to these shifts, the United States must make real investments in high-quality education, at all levels of education.

All Children Get Off to the Right Start with Access to High-Quality Early Education

In a world of AI-driven skill-biased technological change, people with low levels of even basic skills such as reading and math are at higher risk of displacement. On average, children from poor families score far below their peers from higher-income families in early vocabulary and literacy development, in early math, and in the social skills they need to get along well in their classrooms.[54] Studies indicate that kids who start off with deficits in basic skills fail to catch up to peers by later grades.[55] Therefore, it is all the more important that the United States make key investments in getting kids from all income backgrounds off to the right start. To achieve these goals, the United States must catch up to the rest of the world in pre-school enrollment—the United States is ranked 28th out of 38 OECD countries in the share of 4-year-olds enrolled in early education programs.[56]

---

[50] U.S. Department of Education, "A Matter of Equity: Preschool in America," April 2015 (https://www2.ed.gov/documents/early-learning/matter-equity-preschool-america.pdf).
[51] OECD, "Program for International Student Assessment 2012 Results (https://www.oecd.org/unitedstates/PISA-2012-results-US.pdf).
[52] *Ibid.*
[53] The White House, "Remarks of the President to the United States Hispanic Chamber of Commerce," March 2009 (https://www.whitehouse.gov/the-press-office/remarks-president-united-states-hispanic-chamber-commerce).
[54] U.S. Department of Education, "A Matter of Equity: Preschool in America," April 2015 (https://www2.ed.gov/documents/early-learning/matter-equity-preschool-america.pdf).
[55] Connie Juel, "Learning to read and write: A longitudinal study of 54 children from first through fourth grades," *Journal of Educational Psychology* 80.4:437, 1988; David J. Francis, et al., "Developmental lag versus deficit models of reading disability: A longitudinal, individual growth curves analysis," *Journal of Educational Psychology*, 88.1:3, 1996.
[56] OECD, "Education at a Glance: OECD Indicators 2012," Country Note: United States, 2012 (https://www.oecd.org/unitedstates/CN%20-%20United%20States.pdf).

All Students Graduate from High School College- and Career-Ready

Students are much better positioned for jobs that benefit from AI instead of being replaced by it if they graduate from high school with the necessary skills. In the Obama Administration, the U.S. high school graduation rate has reached a record high of 83 percent for the 2014-2015 school year.[57] Too many students, however, are not college- and career-ready when they finish high school. According to the National Assessment of Education Progress, the largest nationally representative and continuing assessment of what America's students know and can do, fewer than 40 percent of graduating students scored at college- and career-ready levels in 2013.[58] A strategy to dramatically speed up school improvement to keep pace with the accelerating rate of change in the global economy should include attracting and retaining the best teachers, making sure all schools have the resources required for success, and holding all students to high standards with rigorous coursework.

Additionally, it will require building on the President's *Computer Science for All* initiative, which seeks to give all students at the K-12 level access to coursework in computing and computational thinking. A bipartisan coalition of governors, mayors, and other public- and private-sector leaders has supported the creation of new standards, courses, and investments in teacher professional development, as well as supplementary extracurricular programs and resources to make this a reality. Further effort is needed to make computer science education available to all children.

The United States has made unprecedented progress ensuring more schools and libraries can access the digital tools to dramatically improve educational outcomes, particularly in delivering technology skills. From 2013 to 2015, 20 million more students gained access to high-speed broadband and wireless in schools, halving the connectivity divide, and the United States is on track to connect 99 percent of students in the near future. But while significant progress has been made, there is still more work to be done to ensure that all schools, libraries, and homes have access to broadband-enabled devices. And educators will need more support and professional development to deliver high-quality learning tailored to their students' needs.

All Americans Have Access to an Affordable Post-Secondary Education that Prepares Them for Good Jobs

Projections show that in the coming years nearly three-quarters of the fastest growing occupations will require at least some postsecondary education beyond high school.[59] Despite these growing needs, state funding for higher education is down by 18 percent on average since the start of the recession, and tuition at four-year public colleges is up by 33 percent since the

---

[57] National Center for Education Statistics, "The Condition of Education: Public High School Graduation Rates," May 2016.
[58] U.S. Department of Education, Institute of Education Sciences, National Center for Education Statistics, National Assessment of Educational Progress (NAEP), 2013 Mathematics and Reading Assessments.
[59] Bureau of Labor Statistics, "Employment Projections—2014-24," December 2015 (http://www.bls.gov/news.release/pdf/ecopro.pdf).

2007-2008 school year.[60] To further improve college access, affordability, and completion, the President has proposed making 2 years of community college free for hard-working students through America's College Promise. If all states participated in America's College Promise, an additional 9 million students could benefit.[61] This could be critical for workers seeking to gain new skills as a response to, or in order to avoid, dislocation from AI-driven automation.

Expand access to training and re-training

A commitment to preparing Americans to adapt to continuous and rapid technological change in the future, whether in AI or other fields, requires pursuing policy changes that would significantly expand the availability of high-quality job training to meet the scale of need; help people more successfully navigate job transitions; and target resources to programs that are producing strong results. But despite the clear challenges facing U.S. workers, the current level of investment in active labor market policies, such as training programs, by the United States is low by both international and historical standards. Through the Workforce Innovation and Opportunity Act—the Federal Government's largest job training investment program—only about 175,000 people are trained per year.[62] As shown above in Figures 9 and 10, while the member countries of the OECD spent, on average, 0.6 percent of GDP on active labor market policies in 2014, spending by the United States was just 0.1 percent of GDP. Relative to the overall economy, the United States now spends less than half of what it did on such programs 30 years ago.

The steps described below should be taken to put the United States ahead of, rather than behind, the curve in these critical areas.

Significantly Expand Availability of Job-Driven Training and Lifelong Learning to Meet the Scale of the Need

Increasing funding for job training by six-fold—which would match spending as a percentage of GDP to Germany, but still leave the United States far behind other European countries—would enable retraining of an additional 2.5 million people per year.[63] Over the last 8 years, the Administration has taken important steps to move in this direction. The Administration has made investments in half of community colleges in the country to create job-driven training programs in fields such as healthcare, information technology, energy and other in-demand fields that have trained nearly 300,000 people so far. The Administration also launched the POWER Initiative, a

---

[60] Center on Budget and Policy Priorities, "Funding Down, Tuition Up: State Cuts to Higher Education Threaten Quality and Affordability at Public Colleges," August 2016 (http://www.cbpp.org/research/state-budget-and-tax/funding-down-tuition-up).
[61] The White House, "Fact Sheet—White House Unveils America's College Promise Proposal: Tuition-Free Community College for Responsible Students," January 2015 (https://www.whitehouse.gov/the-press-office/2015/01/09/fact-sheet-white-house-unveils-america-s-college-promise-proposal-tuitio).
[62] Department of Labor, Employment and Training Administration, 2015.
[63] This assumes $6,000 per person training/reemployment cost, and an increase in Workforce Innovation and Opportunity Act funding from today's $3B to $18B, to match Germany's spending as a fraction of GDP, with all new funding spent on training.

new interagency effort to assist communities negatively impacted by changes in the coal industry and power sector with coordinated Federal economic and workforce-development resources.

Target Resources to Effective Education and Training Programs

Directing funding to training that produces results starts with information about whether programs are placing people into in-demand jobs that pay good salaries. Historically, very little data on the employment outcomes of education and training institutions has existed. Federal and State policymakers should build on Administration initiatives, such as the College Scorecard and the Workforce Innovation and Opportunity Act, by continuing to make investments in collecting and analyzing data on employment and earnings outcomes to hold programs accountable and direct funding to strategies producing results.

Expand Access to Apprenticeships

Job-driven apprenticeships grow the economy and can provide American workers from all backgrounds with the skills and knowledge they need to adapt to a changing economy. Research suggests that apprentices earn a significant premium for their skills—as much as $300,000 more than their peers over a lifetime.[64] The Obama Administration has prioritized expanding apprenticeship programs, and in 2014, called for doubling the number of U.S. registered apprenticeships over the next 5 years. The Administration awarded $175 million to expand apprenticeship in 2015 and announced $50.5 million in grants to support the expansion of new apprenticeship opportunities across the country in all major industry sectors in October 2016.[65]

Strategy #3: Aid Workers in the Transition and Empower Workers to Ensure Broadly Shared Growth

As AI-driven automation changes the economy, empowered workers can be one of the Nation's greatest assets. They can drive and spread innovation, lift consumer demand, and invest in the next generation.[66] This strategy explores how to ensure that workers and job seekers are able to pursue the job opportunities for which they are best qualified, able to bounce back successfully

---

[64] Debbie Reed, et al. An Effectiveness Assessment and Cost-Benefit Analysis of Registered Apprenticeship in 10 States. Mathematica Policy Research, 2012 (https://wdr.doleta.gov/research/FullText_Documents/ETAOP_2012_10.pdf).

[65] The White House, "Fact Sheet: Investing More than $50 Million through ApprenticeshipUSA to Expand Proven Pathways onto the Middle Class." (https://www.dol.gov/sites/default/files/newsroom/newsreleases/WhiteHouseFactSheet-ApprenticeshipUSA-FY2016.pdf)

[66] Ragnhild Balsvik, "Is labor mobility a channel for spillovers from multinationals? Evidence from Norwegian manufacturing," *The Review of Economics and Statistics* 93.1: 285-297, 2011; April M. Franco and Matthew F. Mitchell, "Covenants not to compete, labor mobility, and industry dynamics." *Journal of Economics & Management Strategy* 17.3: 581-606, 2008; Florence Honore, "From Common Ground to Breaking New Ground: Founding Teams' Prior Shared Experience and Start-up Performance." Univ. of Minnesota, 2014; Council of Economic Advisers, "Labor Market Monopsony: Trends Consequences, and Policy Responses," 2016; Council of Economic Advisers, "Worker Voice in a Time of Rising Inequality," 2015; The White House, "Economic Report of the President," Chapter 4. February 2016 (http://go.wh.gov/dM4yPt).

from job loss, and be well-positioned to ensure they receive an appropriate return for their work in the form of rising wages.

Modernize and strengthen the social safety net

Changes to how people work and the dislocation of some workers due to automation heightens the need for a robust safety net to ensure that people can still make ends meet, retrain, and potentially transition careers. That means strengthening critical supports such as unemployment insurance, Medicaid, Supplemental Nutrition Assistance Program (SNAP), and Temporary Assistance for Needy Families (TANF), and putting in place new programs such as wage insurance and emergency aid for families in crisis. It also means exploring whether programs such as Trade Adjustment Assistance should be expanded to help those displaced by automation.

In addition, with the rise of part-time and contingent work, and a more mobile workforce in which individuals do not spend their entire career at a single company, policymakers will need to ensure that workers can access retirement, health care, and other benefits whether or not they get them on the job. For example, the Affordable Care Act expanded eligibility for Medicaid and reformed individual health insurance market to ensure that Americans who do not get coverage on the job can still find affordable, high-quality coverage, while simultaneously introducing reforms to improve coverage for people who are offered coverage at work.

Strengthen Unemployment Insurance

Job displacement is likely to be one of the most serious negative consequences of AI-driven automation, impacting entire industries and communities. Since its inception, unemployment insurance has been a powerful tool to prevent a job loss from hurtling a family into poverty. Last year alone, more than 7 million working Americans relied on the program to get by in tough times. Yet its protections have weakened over time, and today coverage by the program is at its lowest level in at least 50 years.[67] Fewer than one in three unemployed Americans now receive unemployment insurance benefits, and benefits replace a smaller percentage of wages than before for those who do qualify. In 2009, the American Recovery and Reinvestment Act built a foundation for modernizing the program by making $7 billion available to states to expand coverage. Since then, more than 30 states have implemented and sustained important reforms.

The program will need to be further strengthened, as laid out in a proposal to Congress from the President, to ensure that the program can offer a more secure safety net for workers displaced by AI-driven automation and to provide a countervailing force against regional spikes in unemployment. Because workers may be unemployed for longer periods of time as they retrain or shift occupations, benefits should be restored to 26 weeks across the country. The program should also provide up to 52 weeks of additional benefits in states experiencing high levels of unemployment or rapid job loss to dampen the effects of mass layoffs on local economies. This would minimize the chance that these layoffs would lead to broader regional or national recessions. Finally, with fewer than 20 states having sufficient reserves to weather even a single

---

[67] Internal OMB calculations, with unemployment insurance coverage measured as UI claims divided by the number of unemployed individuals.

year of recession, the system's long-term solvency must be ensured so that states are fully prepared for the potential increased cost of benefits stemming from job losses.

Some new tools could also help. Work-sharing programs can help employers hold on to their workers by reducing hours instead of laying them off, with workers whose hours are cut receiving partial unemployment benefits. States could also adopt temporary work-based training programs and allow workers to continue receiving unemployment benefits while participating in on-the-job training.

Experienced workers who lose their jobs and have to start over find themselves, on average, earning wages at least 10 percent less than what they earned in the jobs they lost, and workers with more than 20 years of experience in their prior job face wages that are nearly a quarter less than they had previously been making.[68] For this reason, the President has proposed providing wage insurance to workers who were displaced from jobs if they take a job earning less, replacing up to half of their lost wages. Such a program would help soften the blow of the lower wages some displaced workers would have to accept, and would encourage workers to put their skills back to work quickly so they do not join the ranks of the long-term unemployed or leave the workforce.

These proposals represent steps that can be taken to prepare for a continuation of the job impacts that are already occurring in the economy. If, however, the economic impact of AI is relatively intense or comes on relatively quickly, and if the number of jobs affected approaches the estimates of Frey and Osbourne, then the unemployment insurance system may need to be significantly upgraded to match the magnitude of economic disruption and ensure that displaced workers do not leave the labor force.

Give Workers Improved Guidance to Navigate Job Transitions

With the AI revolution, guidance about how to effectively navigate this transition will be all the more critical. Simple and relatively inexpensive services such as job-search assistance, advice about education and training, and access to labor-market information have been found to be quite effective at helping individuals looking for work find employment more quickly. Evaluations typically find that employment services speed up employment by one to two weeks.[69] Additionally, more intensive counseling services have been shown to increase recipients' earnings and decrease the time they spend unemployed.[70] AI can also be applied, to help workers find information that is best suited to their particular skills and circumstances.

---

[68] The White House, "Fact Sheet: Improving Economic Security by Strengthening and Modernizing the Unemployment Insurance System," Executive Office of the President, January 2016 (https://www.whitehouse.gov/the-press-office/2016/01/16/fact-sheet-improving-economic-security-strengthening-and-modernizing).

[69] Louis Jacobson and Ian Petta, "Measuring the Effect of Public Labor Exchange (PLX) Referrals and Placements in Washington and Oregon," *OWS Occasional Paper 6*, 2000 (https://wdr.doleta.gov/research/FullText_Documents/owsop_2000_06.pdf).

[70] Marios Michaelides, Eileen Poe-Yamagata, Jacob Benus, and Dharmendra Tirumalasetti, "Impact of the Reemployment and Eligibility Assessment (REA) Initiative in Nevada," IMPAQ International, 2012

Strengthen Other Safety Net Programs

Other programs, such as SNAP and TANF, can provide critical safeguards for individuals who have lost their jobs or seen a substantial drop in income, by supplementing low-income households with food and monetary assistance. SNAP has played an important role in lifting millions of Americans out of poverty over the past five decades by providing key nutrition in times of need, with research showing that its benefits go beyond alleviating hunger to improving short-run and long-run health, educational attainment, and economic self-sufficiency.[71]

The TANF program, in conjunction with SNAP, was originally intended to help needy families achieve self-sufficiency. If automation leads to a rise in the number of families needing basic assistance due to growing inequality and poverty, it will be important to have a strong TANF program in place to help especially hard-hit families get back on their feet and work toward self-sufficiency. The President put forward a package of proposals to strengthen TANF, shoring up a system that has eroded over time, which would better support families affected by AI-driven automation.

In addition, the President's budget includes a new $2 billion in funds to test and scale innovative state and local approaches to aid families facing financial crisis. For families on the brink—including those affected by the entry of automation into a sector or occupation—a temporary illness or broken-down car could put them over the edge into a cycle of poverty. The funding would provide families with the emergency help they need to avert or reverse a downward spiral, and then connect those who need it with longer-term assistance so that families are stabilized.

(http://www.impaqint.com/sites/default/files/project-reports/ETAOP_2012_08_REA_Nevada_Follow_up_Report.pdf).

[71] The White House, "Long-Term Benefits of the Supplemental Nutrition Assistance Program," Executive Office of the President, December 2015
(https://www.whitehouse.gov/sites/whitehouse.gov/files/documents/SNAP_report_final_nonembargo.pdf).

> Box 3: Replacing the Current Safety Net with a Universal Basic Income Could Be Counterproductive
>
> *(Excerpt from a speech by CEA Chair Jason Furman, in New York, July 7, 2016)*
>
> Fears of mass job displacement as a result of automation and AI, among other motivations, have led some to propose deep changes to the structure of government assistance. One of the more common proposals has been to replace some or all of the current social safety net with a universal basic income (UBI): providing a regular, unconditional cash grant to every man, woman, and child in the United States, instead of, say, Temporary Assistance to Needy Families (TANF), the Supplemental Nutrition Assistance Program (SNAP), or Medicaid.
>
> While the exact contours of various UBI proposals differ, the idea has been put forward from the right by Charles Murray (2006), the left by Andy Stern and Lee Kravitz (2016), and has been a staple of some technologists' policy vision for the future (Rhodes, Krisiloff, and Altman 2016). The different proposals have different motivations, including real and perceived deficiencies in the current social safety net, the belief in a simpler and more efficient system, and also the premise that we need to change our policies to deal with the changes that will be unleashed by AI and automation more broadly.
>
> The issue is not that automation will render the vast majority of the population unemployable. Instead, it is that workers will either lack the skills or the ability to successfully match with the good, high paying jobs created by automation. While a market economy will do much of the work to match workers with new job opportunities, it does not always do so successfully, as we have seen in the past half-century. We should not advance a policy that is premised on giving up on the possibility of workers' remaining employed. Instead, our goal should be first and foremost to foster the skills, training, job search assistance, and other labor market institutions to make sure people can get into jobs, which would much more directly address the employment issues raised by AI than would UBI.

Increase wages, competition, and worker bargaining power

As the Council of Economic Advisers recently discussed in its report on labor market monopsony,[72] there is growing concern about a general reduction in competition among firms for workers and a commensurate shift in the balance of bargaining power toward employers. Market concentration resulting from the development of AI has the potential to worsen these trends. A displacement of workers from industries being disrupted by AI-driven automation would also create slack in the labor market in the short to medium-run, which is likely to depress wages.

---

[72] Council of Economic Advisers, "Issue Brief: Labor Market Monopsony: Trends, Consequences, and Policy Responses," October 2016 (https://www.whitehouse.gov/sites/default/files/page/files/20161025_monopsony_labor_mrkt_cea.pdf).

Fortunately, there are several approaches described below that can counterbalance these trends and boost Americans' wages and working conditions.

Raise the Minimum Wage

The minimum wage plays a critical role in reducing inequality, increasing consumption, and strengthening the workforce. Raising it could lift at least 4.6 million people out of poverty.[73] Adjusted for inflation, the value of the minimum wage has fallen by nearly a quarter from its peak value in 1968 and is about one-fifth less than it was when President Reagan took office. While Congress has not acted, 22 states and the District of Columbia enacted legislation raising their minimum wage since that time—including the direct passage of new increases by residents of four states (Arizona, Colorado, Maine, and Washington) in the November 2016 elections.

Modernize Overtime and Spread Work

Offering overtime is one of the single most important steps to help grow middle-class wages and spread jobs to more workers. In May 2016, the Department of Labor finalized revisions to its regulations that would extend overtime protections to 4.2 million more Americans and boost wages for workers by $12 billion over the next decade.[74]

Strengthen Unions, Worker Voice, and Bargaining Power

Growing and sustaining the middle class requires strong labor unions. Labor unions help to build the middle class and have been critical in restoring the link between hard work and opportunity so the benefits of economic growth can be more broadly shared.[75] Unions have been at the forefront of establishing the 40-hour work week and the weekend, eliminating child labor laws, and establishing fair benefits and decent wages. Policymakers should explore ways to empower worker voice in the workplace through strengthening protections for organizing and creating new and innovative ways for workers to make their voices heard.

Protect Wages

Given the unique transformation that may be brought by AI, policymakers may also want to consider whether additional wage protections are needed for low- and middle-skilled workers if automation further "hollows out" middle-skill jobs. One of the most powerful upward pressures on wages is a tight labor market, as demand for labor can drive up wages.

---

[73] Arindrajit Dube, "Minimum Wages and the Distribution of Family Incomes," 2014 (http://sites.utexas.edu/chasp/files/2015/04/MinimumWagesandDistributionofFamilyIncomes.pdf)
[74] The White House, "Fact Sheet: Growing Middle Class Paychecks and Helping Working Families Get Ahead by Expanding Overtime Pay," Executive Office of the President, May 2016 (https://www.whitehouse.gov/the-press-office/2016/05/17/fact-sheet-growing-middle-class-paychecks-and-helping-working-families-0). This rule was challenged in court and is currently subject to an injunction.
[75] Council of Economic Advisers, "Issue Brief: Worker Voice in a Time of Rising Inequality," October 2015 (https://www.whitehouse.gov/sites/default/files/page/files/cea_worker_voice_issue_brief.pdf).

Identify strategies to address differential geographic impact

Automation will happen more quickly in some places than in others, because of local policies, access to capital, innovative thinkers, the skill set of the workforce, proximity to urban centers, the culture of a place, and myriad other reasons. This has the potential to further exacerbate geographic disparities in income and wealth. Many of the places that are already grappling with structural changes in the economy, overall economic shifts, and poverty—and that therefore seem left behind from today's economy—may fall even further behind, cementing an already-present divide.

Below are two ways to address the uneven geographic impact: reduce the geographic barriers to work and pursue "place-based" solutions.

Reduce Geographic Barriers to Work

Geographic inequality can be reduced if workers can move to areas with more opportunity. If the new jobs spawned by AI-driven automation continue the trend of urbanization, though, a lack of affordable housing could make it difficult for lower-income families to access them. Over the past 3 decades, a growing number of barriers—including zoning, land-use regulations, and lengthy development approval processes—have made it increasingly difficult for housing markets to respond to growing demand by increasing the supply of housing. Reducing these barriers to affordable housing, expanding broadband access in poor and rural areas, and improving public transit would all serve to reduce geographic barriers to work.

Another barrier is occupational licensing. Nearly one-quarter of all U.S. workers need a government license to do their jobs.[76] While licensing can offer important health and safety protections to consumers, as well as benefits to workers, the current system often requires unnecessary training, lengthy delays, or high fees. Research shows that licensing can not only reduce total employment in licensed professions, but also that unlicensed workers earn roughly 7 percent lower wages than licensed workers with similar levels of education and experience.[77] In addition, the patchwork of state-by-state licensing rules leads to dramatically different requirements for the same occupations depending on the state in which one lives, potentially burdening workers who aim to move across state lines frequently. Best practices in licensing could allow states, working together or individually, to safeguard the well-being of consumers while maintaining a modernized regulatory system that meets the needs of workers and businesses. For example, groups of states could harmonize regulatory requirements as much as possible, and where appropriate enter into inter-state compacts that recognize licenses from other states to increase the mobility of skilled workers.[78]

---

[76] The White House, "Occupational Licensing: A Framework for Policymakers," July 2015 (https://www.whitehouse.gov/sites/default/files/docs/licensing_report_final_nonembargo.pdf).
[77] Bureau of Labor Statistics. CEA calculations (https://www.whitehouse.gov/blog/2016/06/17/new-data-show-roughly-one-quarter-us-workers-hold-occupational-license).
[78] The White House, "Occupational Licensing: A Framework for Policymakers," July 2015 (https://www.whitehouse.gov/sites/default/files/docs/licensing_report_final_nonembargo.pdf).

Pursue Place-Based Solutions

To the extent that AI will differentially affect certain cities, policies should be appropriately targeted and can build on and learn from existing Administration policies. The Administration has launched a series of "place-based" initiatives, often focused on economically challenged areas, that empower local leaders in participating communities by helping to connect them with the resources they need to enact their own homegrown solutions for the challenges facing their communities.

Initiatives such as Choice Neighborhoods and Promise Zones take a big-picture approach to development. Their work helps improve communities' access to and delivery of a wide range of services and activities, from building housing and creating jobs, to supporting after-school programs and improving the health of local residents. Agencies across the Federal Government have worked together to cut through red tape and expand opportunity for the people they serve.

Another example, TechHire, is a national initiative to create pathways for more Americans to access well-paying technology jobs and expand local technology sectors in communities across the country. Originally announced by President Obama in 2015, TechHire enables employers to fill entry-level, career-path, skilled technology jobs, by hiring trained job seekers with the ability to do the job—but who are overlooked by typical hiring practices or underrepresented in the IT field.

Modernize tax policy

Tax policy plays a critical role in combating inequality, including income inequality that may be exacerbated by changes in employment from AI-based automation. A progressive tax system helps ensure that the benefits of economic growth are broadly shared, pushing back against increased inequality in pre-tax income. Progressive taxation is critical for raising adequate revenue to fund national security and domestic priorities, including supporting and retraining workers who may be harmed by increased automation, and will only grow more important if outsized gains continue to accrue at the top while other workers are left struggling. It is also critical to maintain and strengthen tax credits that encourage and reward work while helping families make ends meet, such as the Earned Income Tax Credit and Child Tax Credit.

Advanced AI systems could reinforce trends of national income shifting from labor to capital, as discussed above. With investment income heavily concentrated among high-income individuals, this shift could significantly exacerbate the rise in income inequality seen over the past few decades, absent an appropriate policy response. Taxing capital can be a highly progressive form of taxation, yet under the current tax system, individuals' capital income currently enjoys lower tax rates than labor income, and often goes untaxed. President Obama has proposed reforming capital taxation and raising revenue by ensuring that inherited assets are subject to capital gains tax (ending so-called "stepped up basis"); increasing the top rate on capital gains and dividends for high-income households to 28 percent, the rate under President Reagan; and increasing the estate tax by restoring its 2009 parameters and closing loopholes. Some experts have proposed other reforms, including taxing capital gains on a mark-to-market basis and further reforming taxes on wealth transfers.

Preparing for all contingencies

If job displacements from AI are considerably beyond the patterns of technological change previously observed in economic history, a more aggressive policy response would likely be needed, with policymakers potentially exploring countervailing job creation strategies, new training supports, a more robust safety net, or additional strategies to combat inequality.

Conclusion

Responding to the economic effects of AI-driven automation will be a significant policy challenge for the next Administration and its successors. AI has already begun to transform the American workplace, changing the types of jobs available and the skills that workers need to thrive. All Americans should have the opportunity to participate in addressing these challenges, whether as students, workers, managers, or technical leaders, or simply as citizens with a voice in the policy debate.

AI raises many new policy questions, which should be continued topics for discussion and consideration by future Administrations, Congress, the private sector, and the public. Continued engagement among government, industry, technical and policy experts, and the public should play an important role in moving the Nation toward policies that create broadly shared prosperity, unlock the creative potential of American companies and workers, and ensure the Nation's continued leadership in the creation and use of AI.

References

Daron Acemoglu, "Directed Technological Change, *Review of Economic Studies* 69(4): 781-809, 2003 (http://restud.oxfordjournals.org/content/69/4/781.short).

_____, NBER reporter article, 2002.

_____, "Technical change, inequality, and the labor market," *Journal of economic literature* 40(1): 7-72, 2002.

Daron Acemoglu and David Autor, "Skills, tasks and technologies: Implications for employment and earnings," 2011, *Handbook of labor economics* 4(2011): 1043-171, (http://economics.mit.edu/files/5571).

Ajay Agrawal, Joshua Gans, and Avi Goldfarb, "The Simple Economics of Machine Intelligence," *Harvard Business Review,* November 17 2016 (https://hbr.org/2016/11/the-simple-economics-of-machine-intelligence).

Melanie Arntz, Terry Gregory, and Ulrich Zierahn, "The Risk of Automation for Jobs in OECD Countries: A Comparative Analysis," OECD Social, Employment and Migration Working Papers No. 189, 2016 (http://www.oecd-ilibrary.org/docserver/download/5jlz9h56dvq7-en.pdf?expires=1480994298&id=id&accname=guest&checksum=6DC4B241A91EE860DC391585FF43C51C).

David H. Autor and David Dorn, "The Growth of Low-Skill Service Jobs and the Polarization of the US Labor Market," *American Economic Review* 103(5): 1553-97, 2013 (http://www.ddorn.net/papers/Autor-Dorn-LowSkillServices-Polarization.pdf).

David H Autor, David Dorn, and Gordon H. Hanson, "The China Syndrome: Local Labor Market Effects of Import Competition in the United States," *American Economic Review* 103(6): 2121-68, 2013 (http://gps.ucsd.edu/_files/faculty/hanson/hanson_publication_it_china.pdf).

David H. Autor, Frank Levy, and Richard J. Murnane, "The Skill Content of Recent Technological Change: An Empirical Exploration." *Quarterly Journal of Economics* 118(4): 1279-1333, 2003.

Ragnhild Balsvik, "Is labor mobility a channel for spillovers from multinationals? Evidence from Norwegian manufacturing," *The Review of Economics and Statistics* 93.1: 285-297, 2011.

Susanto Basu, John G. Fernald, and Matthew D. Shapiro, "Productivity growth in the 1990s: technology, utilization, or adjustment?" *Carnegie-Rochester Conference Series on Public Policy* 55(1): 117-65, 2001 (https://ideas.repec.org/a/eee/crcspp/v55y2001i1p117-165.html).

Amy Bernstein and Anand Raman, The Great Decoupling: An Interview with Erik Brynjolfsson and Andrew McAfee, *Harvard Business Review*, June 2015 (https://hbr.org/2015/06/the-great-decoupling).

Erik Brynjolfsson and Andrew McAfee, *The Second Machine Age: Work, Progress, and Prosperity in a Time of Brilliant Technologies,* WW Norton & Company, 2014.

_____, CEA calculations (https://www.whitehouse.gov/blog/2016/06/17/new-data-show-roughly-one-quarter-us-workers-hold-occupational-license).

John W. Budd, *The Thought of Work*, Cornell University Press, 2011.

Bureau of Labor Statistics, Civilian Unemployment Rate, 1948-2016.

_____, "Employment Projections—2014-24," December 2015 (http://www.bls.gov/news.release/pdf/ecopro.pdf).

_____, "Employment Projections: Employment by major industry sector," December 2015 (http://www.bls.gov/emp/ep_table_201.htm).

Center on Budget and Policy Priorities, "Funding Down, Tuition Up: State Cuts to Higher Education Threaten Quality and Affordability at Public Colleges," August 2016 (http://www.cbpp.org/research/state-budget-and-tax/funding-down-tuition-up).

Kerwin Kofi Charles, Erik Hurst, and Matthew J. Notowidigdo, "Housing Booms, Manufacturing Decline, and Labor Market Outcomes." Working Paper, 2016 (http://faculty.wcas.northwestern.edu/noto/research/CHN_manuf_decline_housing_booms_mar2016.pdf).

Ajay Chaudry, "The Case for Early Education in the Emerging Economy," *Roosevelt Institute-The New American Economy's Learning* Series, August 2016, p. 3 (http://rooseveltinstitute.org/wp-content/uploads/2016/08/The-Case-for-Early-Education-in-the-Emerging-Economy.pdf).

Council of Economic Advisers, "Labor Market Monopsony: Trends Consequences, and Policy Responses," 2016.

_____, "The Economic Record of the Obama Administration: Reforming the Health Care System," December 2016 (https://www.whitehouse.gov/sites/default/files/page/files/20161213_cea_record_healh_care_reform.pdf).

_____, "Worker Voice in a Time of Rising Inequality," 2015.

Cyber Grand Challenge (https://www.cybergrandchallenge.com).

Patricia A. Daly, "Agricultural Employment: Has the Decline Ended?" *Monthly Labor Review* November 1981 (http://www.bls.gov/opub/mlr/1981/11/art2full.pdf).

Paul Daugherty and Mark Purdy. "Why AI is the Future of Growth." 2016 (https://www.accenture.com/t20161031T154852__w__/us-en/_acnmedia/PDF-33/Accenture-Why-AI-is-the-Future-of-Growth.PDF#zoom=50).

Steven J. Davis and Till Von Wachter, "Recessions and the Costs of Job Loss," Brookings Papers on Economic Activity, Economic Studies Program, The Brookings Institution, 43(2): 1-72, 2011 (http://www.econ.ucla.edu/workshops/papers/Monetary/Recessions%20and%20the%20Costs%20of%20Job%20Loss%20with%20Appendix.pdf).

Arindrajit Dube, "Minimum Wages and the Distribution of Family Incomes," 2014 (http://sites.utexas.edu/chasp/files/2015/04/MinimumWagesandDistributionofFamilyIncomes.pdf).

Ed Felten and Terah Lyons, "Public Input and Next Steps on the Future of Artificial Intelligence," *Medium*, September 6 2016 (https://medium.com/@USCTO/public-input-and-next-steps-on-the-future-of-artificialintelligence-458b82059fc3).

David J. Francis, et al., "Developmental lag versus deficit models of reading disability: A longitudinal, individual growth curves analysis," *Journal of Educational Psychology,* 88.1:3, 1996.

April M. Franco and Matthew F. Mitchell, "Covenants not to compete, labor mobility, and industry dynamics." *Journal of Economics & Management Strategy* 17.3: 581-606, 2008.

Carl Frey and Michael Osborne, "The Future of Employment: How Susceptible are Jobs to Computerization," Oxford University, 2013 (http://www.oxfordmartin.ox.ac.uk/downloads/academic/The_Future_of_Employment.pdf).

Jason Furman, "Productivity Growth in the Advanced Economies: The Past, the Present, and Lessons for the Future." Speech at the Peterson Institute for International Economics, Washington, July 9 2015 (https://www.whitehouse.gov/sites/default/files/docs/20150709_productivity_advanced_economies_piie.pdf).

_____, "Is This Time Different? The Opportunities and Challenges of Artificial Intelligence." Remarks at AI Now: The Social and Economic Implications of Artificial Intelligence Technologies in the Near Term. New York University, New York, July 7, 2016. (https://www.whitehouse.gov/sites/default/files/page/files/20160707_cea_ai_furman.pdf).

Claudia Goldin and Lawrence F. Katz, *The Race between Education and Technology*, 2008.

Georg Graetz and Guy Michaels, "Robots at Work," *CEPR Discussion Paper No. DP10477*, March 2015 (http://papers.ssrn.com/sol3/papers.cfm?abstract_id=2575781).

Jonathan V. Hall and Alan B. Krueger, "An Analysis of the Labor Market for Uber's Driver-Partners in the United States," 2016, (http://www.nber.org/papers/w22843.pdf).

Seth D. Harris and Alan B. Krueger, "A Proposal for Modernizing Labor Laws for Twenty-First-Century Work: The 'Independent Worker,'" 2015, (http://www.hamiltonproject.org/assets/files/modernizing_labor_laws_for_twenty_first_century_work_krueger_harris.pdf).

Florence Honore, "From Common Ground to Breaking New Ground: Founding Teams' Prior Shared Experience and Start-up Performance." Univ. of Minnesota, 2014.

David Hounshell, *From the American system to mass production, 1800-1932: The development of manufacturing technology in the United States*, Baltimore: JHU Press, 1985.

Vivian Hunt, Dennis Layton, and Sara Prince, "Why Diversity Matters," McKinsey & Company, 2015 (http://www.mckinsey.com/business-functions/organization/our-insights/why-diversity-matters).

Louis Jacobson and Ian Petta, "Measuring the Effect of Public Labor Exchange (PLX) Referrals and Placements in Washington and Oregon," *OWS Occasional Paper 6*, 2000 (https://wdr.doleta.gov/research/FullText_Documents/owsop_2000_06.pdf).

Nir Jaimovich and Henry E. Siu, "The Trend is the Cycle: Job Polarization and Jobless Recoveries." NBER Working Paper No. 18334, 2012 (http://www.nber.org/papers/w18334.pdf).

John A. James and Jonathan S. Skinner, "The Resolution of the Labor-Scarcity Paradox," *The Journal of Economic History,* 45(3): 513-40, 1985 (http://www.jstor.org/stable/2121750?seq=1#page_scan_tab_contents).

Aparna Joshi and Hyntak Roh, "The Role of Context in Work Team Diversity Research: A Meta-Analytic Review," *Academy of Management Journal* 52: 599-627, 2009.

Connie Juel, "Learning to read and write: A longitudinal study of 54 children from first through fourth grades," *Journal of Educational Psychology* 80.4:437, 1988.

Lawrence F. Katz and Kevin M. Murphy, "Changes in Relative Wages, 1963-1987: Supply and Demand Factors," *Quarterly Journal of Economics*, 107(Feb): 35-78, 1992.

John M. Keynes, "Economic Possibilities for our Grandchildren." In *Essays in Persuasion*, New York: W.W.Norton & Co., pp. 358-373, 1930. (http://www.econ.yale.edu/smith/econ116a/keynes1.pdf).

Marios Michaelides, Eileen Poe-Yamagata, Jacob Benus, and Dharmendra Tirumalasetti, "Impact of the Reemployment and Eligibility Assessment (REA) Initiative in Nevada," IMPAQ International, 2012 (http://www.impaqint.com/sites/default/files/project-reports/ETAOP_2012_08_REA_Nevada_Follow_up_Report.pdf).

Joel Mokyr, "Technological inertia in Economic History." *The Journal of Economic History*, 52(2): 325-38, 1992 (http://www.jstor.org/stable/2123111?seq=1#page_scan_tab_contents).

National Center for Education Statistics, "The Condition of Education: Public High School Graduation Rates," May 2016.

National Science and Technology Council, *National Artificial Intelligence Research and Development Strategic Plan*, October 2016 (https://www.nitrd.gov/PUBS/national_ai_rd_strategic_plan.pdf).

OECD, "Labour market programmes: expenditure and participants", *OECD Employment and Labour Market Statistics* (database), 2016 (http://stats.oecd.org/viewhtml.aspx?datasetcode=LMPEXP&lang=en#).

_____, Program for International Student Assessment 2012 Results (https://www.oecd.org/unitedstates/PISA-2012-results-US.pdf)

Debbie Reed, et al. An Effectiveness Assessment and Cost-Benefit Analysis of Registered Apprenticeship in 10 States. Mathematica Policy Research, 2012 (https://wdr.doleta.gov/research/FullText_Documents/ETAOP_2012_10.pdf).

Klaus Schwab, The Fourth Industrial Revolution: what it means, how to respond, World Economic Forum, January 2016 (https://www.weforum.org/agenda/2016/01/the-fourth-industrial-revolution-what-it-means-and-how-to-respond/).

USDA Economic Research Service, "Table 1. Indicies of farm output, input, and total factor productivity for the United States, 1948-2013" (https://www.ers.usda.gov/data-products/agricultural-productivity-in-the-us/).

U.S. Department of Education, Institute of Education Sciences, National Center for Education Statistics, National Assessment of Educational Progress (NAEP), 2013 Mathematics and Reading Assessments.

_____, "A Matter of Equity: Preschool in America," April 2015 (https://www2.ed.gov/documents/early-learning/matter-equity-preschool-america.pdf).

Bruce Western and Jake Rosenfeld, "Unions, norms, and the rise in US wage inequality," *American Sociological Review* 76(4): 513-37, 2011.

The White House, *Big Data: Seizing Opportunities, Preserving Values,* May 2014 (https://www.whitehouse.gov/sites/default/files/docs/big_data_privacy_report_may_1_2014.pdf).

_____, "Economic Report of the President," February 2016 (http://go.wh.gov/dM4yPt).

_____, "Fact Sheet: Growing Middle Class Paychecks and Helping Working Families Get Ahead by Expanding Overtime Pay," Executive Office of the President, May 2016 (https://www.whitehouse.gov/the-press-office/2016/05/17/fact-sheet-growing-middle-class-paychecks-and-helping-working-families-0).

_____, "Fact Sheet: Improving Economic Security by Strengthening and Modernizing the Unemployment Insurance System," Executive Office of the President, January 2016 (https://www.whitehouse.gov/the-press-office/2016/01/16/fact-sheet-improving-economic-security-strengthening-and-modernizing).

_____, "Fact Sheet: Investing More than $50 Million through ApprenticeshipUSA to Expand Proven Pathways onto the Middle Class." (https://www.dol.gov/sites/default/files/newsroom/newsreleases/WhiteHouseFactSheet-ApprenticeshipUSA-FY2016.pdf)

_____, "Fact Sheet—White House Unveils America's College Promise Proposal: Tuition-Free Community College for Responsible Students," January 2015 (https://www.whitehouse.gov/the-press-office/2015/01/09/fact-sheet-white-house-unveils-america-s-college-promise-proposal-tuitio).

_____, "Long-Term Benefits of the Supplemental Nutrition Assistance Program," Executive Office of the President, December 2015 (https://www.whitehouse.gov/sites/whitehouse.gov/files/documents/SNAP_report_final_nonembargo.pdf).

_____, "Occupational Licensing: A Framework for Policymakers," July 2015 (https://www.whitehouse.gov/sites/default/files/docs/licensing_report_final_nonembargo.pdf).

_____, "Preparing for the Future of Artificial Intelligence," October 2016 (https://www.whitehouse.gov/sites/default/files/whitehouse_files/microsites/ostp/NSTC/preparing_for_the_future_of_ai.pdf).

_____, "Remarks of the President to the United States Hispanic Chamber of Commerce," March 2009 (https://www.whitehouse.gov/the-press-office/remarks-president-united-states-hispanic-chamber-commerce).

www.ingramcontent.com/pod-product-compliance
Lightning Source LLC
Chambersburg PA
CBHW081126180526
45170CB00008B/3023